Glaciers
Second Edition

Glaciers are among the most beautiful natural wonders on Earth, but for most of us the least known and understood. This new edition describes how glaciers grow and decay, how they move and how they influence human civilization. Today, covering a tenth of the Earth's surface, glacier ice has shaped the landscape over millions of years by scouring away rocks, transporting and depositing debris far from its source. Glacier meltwater drives turbines and irrigates deserts, yields mineral-rich soils and has left us a wealth of valuable sand and gravel. However, glaciers also threaten human property and life. Our future is indirectly bound up with the fate of glaciers and their influence on global climate and sea level.

A lively running text develops these themes and is supported by over 200 stunning photographs, taking us from the High-Arctic through North America, Europe, Asia, Africa, New Zealand and South America to the Antarctic. It builds on the highly acclaimed first edition, including new chapters on Antarctica and Earth's ice age record, as well as many new photographs, now all in full colour.

MICHAEL HAMBREY is Director of the Centre for Glaciology and Professor of Glaciology at the University of Wales, Aberystwyth. He began his academic career investigating glaciers in Norway, the Swiss Alps and the Canadian Arctic, and has since travelled the world to become a highly respected glaciologist.

JÜRG ALEAN teaches geography at Kantonsschule Zürcher Unterland in Bülach, Switzerland, from where he regularly leads student field camps to study glaciers. His early research concentrated on the hazards of ice avalanches, and he regularly travels around the world to work on glaciers.

Glaciers

Second Edition

Michael Hambrey

University of Wales, Aberystwyth

Jürg Alean

Kantonsschule Zürcher Unterland, Bülach

CAMBRIDGE
UNIVERSITY PRESS

PUBLISHED BY THE PRESS SYNDICATE OF THE UNIVERSITY OF CAMBRIDGE
The Pitt Building, Trumpington Street, Cambridge, United Kingdom

CAMBRIDGE UNIVERSITY PRESS
The Edinburgh Building, Cambridge CB2 2RU, UK
40 West 20th Street, New York, NY 10011-4211, USA
477 Williamstown Road, Port Melbourne, VIC 3207, Australia
Ruiz de Alarcón 13, 28014 Madrid, Spain
Dock House, The Waterfront, Cape Town 8001, South Africa

http://www.cambridge.org

First published 2004

Printed in the United Kingdom at the University Press, Cambridge

Typeface Quadraat Regular 11.5/16pt.   System QuarkXPress® [SE]

A catalogue record for this book is available from the British Library

Library of Congress Cataloguing in Publication data

Glaciers/Michael Hambrey and Jürg Alean – 2nd ed.
    p.   cm.
Includes bibliographical references (p.).
ISBN 0 521 82808 2 (hardback)
1. Glaciers.   I. Alean, Jürg.   II. Title.

GB2403.2.H36 2004
551.31′2–dc22   2004043585

ISBN 0 521 82808 2 hardback

The publisher has used its best endeavours to ensure that the URLs for external
websites referred to in this book are correct and active at the time of going to press.
However, the publisher has no responsibility for the websites and can make no
guarantee that a site will remain live or that the content is or will remain appropriate.

Frontispiece. Mer de Glace from
Montenvert, near Chamonix, France; one
of the first glaciers to be studied by
nineteenth-century scientists and now a
popular destination for tourists and
climbers.

Out of whose womb came the ice?
And the hoary frost of heaven,
Who hath gendered it?
The waters are as with a stone,
And the face of the deep is frozen.

Job 38:29–30

The world's highest peak, Mount Everest (8848 metres) on the left, feeds the Khumbu Icefall and then the Khumbu Glacier which flows beneath Nuptse (7861 metres) towards the right. This route represents the main approach to Everest from the south, and is well displayed from this viewpoint of Kalar Patar (5545 metres).

The legacy of glaciation in the area
between the lakes of Zugersee and
Zürichsee, Switzerland is manifested by
drumlins. The ice-streamlined
depositional landforms contain glacial
deposits that provide an important and
reliable aquifer, supplying the nearby
communities with clean drinking water.

Contents

The decaying, pinnacled snout of Wright
Lower Glacier in the polar desert area of
the Dry Valleys of Antarctica.

Preface

Glaciers are one of the most beautiful and fascinating elements of nature. Slowly they creep and slide from mountain regions to the lowlands and cover huge areas of the Polar Regions. Over millions of years glaciers have been shaping landscapes by scouring rocks, and transporting and depositing debris far from its source. In so doing, they have created some of the finest landscapes on Earth. Glaciers provide meltwater that drives turbines and irrigates deserts, furnish material for the development of fertile soils and leave us a rich legacy of sand and gravel of considerable economic value. In contrast to these benefits, glaciers can also destroy human property and take people's lives through ice avalanches and lake-outburst floods.

As glaciologists, we have attempted to understand some of the infinite varieties of glacial phenomena. We have lived for months at a time on, or adjacent to, glaciers and have seen them in all their moods. They have often presented a benign appearance, as on a calm sunny day, when travelling over them has been safer than walking a city street. At other times, such as when blinding blizzards have obliterated our paths and the snow has treacherously hidden crevasses, glaciers have made us wish for the security of home. Yet time and again we have been drawn back to glaciers, eager to absorb their natural beauty as well as to gain a better appreciation of how they behave and to contribute to the science of glaciology.

We have been fortunate to visit glaciers in many parts of the world – from the High-Arctic, through temperate and tropical regimes where they occur only in the highest mountains, to deep in the heart of the Antarctic Ice Sheet. Therefore, in this book we wish to share our appreciation of glaciers, and take the reader, at least in imagination, on excursions to remote lands and to regions close to human occupation.

This volume is the second, revised and expanded edition of *Glaciers*, first published by Cambridge University Press in 1992. The text has been fully revised to take account of the substantial developments in the discipline of glaciology over the past decade. We also provide a number of additional chapters: on Antarctica, on living and travelling on glaciers; on the record of glaciation through the Earth's history; and to look more closely at the future prospects of glaciers.

Firstly, we set the scene by outlining the global context of glaciers, noting briefly several characteristics shared by most glaciers, and commenting on why the study of glaciers is important (Chapter 1). Glaciers come in all shapes and sizes, from a small patch of ice a few hundred metres across to the huge ice sheets that almost totally bury the continent of Antarctica and the island of Greenland, and the many types are described in Chapter 2. The processes that lead to the creation of glaciers, their replenishment and survival are described in Chapter 3. Glaciers are not static: they undergo major fluctuations in response to increases and decreases in temperature and the supply of snow (Chapter 4). Glaciers also flow like thick porridge and slide on their beds, as outlined Chapter 5. This chapter also describes the various structures that result from flow, as well as some of the spectacular events associated with unusual flow behaviour. Glaciers commonly transport large amounts of debris; sometimes they may be completely debris-covered and lack the aesthetic appeal of clean ones (Chapter 6). Most glaciers are intimately associated with meltwater that can provide just as many features and produces just as many landforms as the glaciers themselves (Chapter 7).

Having explored the processes associated with glaciers, we turn, in Chapter 8, to the region that accounts for 91% of the world's ice, Antarctica. We feel that this new chapter does justice to the huge continent of ice centred over the South Pole, which indirectly influences all our lives. On a quite different note, the violent and often disastrous interaction between glaciers and volcanoes forms the basis of Chapter 9. If this mix can create sudden changes to the landscape, even more important ones happen on a slower time scale.

These changes involve glacial erosion and deposition, the many products of which are revealed once ice has receded, giving glaciated landscapes their distinctive character and appeal (Chapter 10).

Next we consider the impact of glaciers on the biosphere. In Chapter 11 we consider how some animals and plants have adapted to severe climatic conditions. Then we consider the benefits of glaciers to humans, such as water and energy supply, tourism and the value of the sedimentary products of glaciers (Chapter 12). We describe the principal hazards linked to glaciers, and describe some of the major catastrophes that have occurred, such as avalanches and floods (Chapter 13). In Chapter 14 we deal with living and travelling on glaciers, in order to explain how scientists go about their research work in such regions; activities range from large-scale government-run research programmes to those of the self-contained independent traveller, while transport ranges from large transport aircraft to foot-slogging.

The present-day extent of ice is placed in the context of geological time in Chapter 15. Here we describe how to identify evidence of past glaciation, describe briefly the ancient glacial record and give examples of some of the major international projects that have helped shape our understanding of the response of glaciers to climatic change.

The book closes (Chapter 16) with a brief review of the future prospects of glaciers, summarizing the evidence for dramatic reduction in ice cover in many parts of the world in response to human-induced climatic warming, but balancing this with what we know of ice sheet stability in Antarctica. Human civilization is undertaking a potentially dangerous 'experiment' by pumping greenhouse gases into the atmosphere. With reference to predicted melting of glaciers and the resulting rise in sea level, we argue that urgent steps are needed to reduce emissions.

To assist the reader, we have highlighted technical terms in bold print. These terms are included in a Glossary at the end of the book.

Nearly all photographs in this book are new to this edition. As before, the majority have been taken by ourselves, and have been selected in order to demonstrate some of the many facets of glacier behaviour. These are supplemented by striking satellite images, especially those made available by NASA on the Internet, and several of other images taken by friends and colleagues. A number of line drawings, maps and tables are included to further illustrate specific topics.

In summary, our aim is to describe and explain glaciers in all their variety, as well as the landscapes they are creating and are still creating. If we succeed in conveying to our readers the beauty and importance of glaciers, we will have fulfilled our purpose.

Acknowledgements

Michael Hambrey

I wish to thank first of all my parents who, perhaps unwittingly, were the first to foster my interest in glaciated landscapes through visits to areas like the English Lake District and North Wales. I am deeply indebted to Wilfred Theakstone of Manchester University, who first inspired my interest in glaciers and guided me through my PhD studies on a Norwegian glacier between 1970 and 1973; Geoffrey Milnes, then of the Swiss Federal Institute of Technology in Zürich, for the opportunity to work with him on Alpine glaciers from 1974 to 1977; the late Fritz Müller of the same institution for opportunities to work on glaciers on Axel Heiberg Island in the Canadian Arctic in 1975; the late Brian Harland, of Cambridge University, for participation in several geological expeditions to Svalbard in the High-Arctic between 1977 and 1983; Martin Sharp, also of Cambridge University, for the opportunity to work on an Alaskan glacier in 1986; Peter Barrett of the Victoria University of Wellington for invitations to join scientifically exciting drilling projects in Antarctica in 1986 and 1999; Niels Henriksen for facilitating fieldwork in East Greenland with the then Geological Survey of Greenland in 1988; Dieter Fütterer, Werner Ehrmann, Gerhard Kuhn and Victor Smetacek of the Alfred Wegener Institute for Polar and Marine Research in Bremerhaven, Germany, for the opportunity to pursue collaborative Antarctic research on FS *Polarstern* to Antarctica (1991); Barrie McKelvey of the University of New England, Armidale, Australia for the invitation to work with him in the remote Prince Charles Mountains in Antarctica, and which entailed flying around much of the edge of the East Antarctic Ice-Sheet (1994); Peter Webb and David Harwood for the opportunity to work with them in the central Transantarctic Mountains in latitude 85°S (1995); Neil Glasser of the University of Wales, Aberystwyth for the chance to work with him in Chilean Patagonia (2000); Sean Fitzsimons of the University of

Otago for the invitation to study glaciers in the Dry Valleys, Antarctica (2001); John Smellie of the British Antarctic Survey for the opportunity to undertake collaborative work in the northern Antarctic Peninsula region, and the officers and crew of the Royal Navy icebreaker HMS *Endurance* which supported us (2002); and John Reynolds and Shaun Richardson of Reynolds Geo-Sciences, North Wales, and Marco Zapata and colleagues of INRENA, Huaraz for introducing me to hazardous glaciers in Peru (2002).

It is a pleasure to acknowledge the stimulating discussions and companionship of many university colleagues, logistical supporters and research students, in addition to the above, who have participated in, and helped organize field programmes in Greenland (1984, 1985), Svalbard (1992–2001), Peru (2002), Nepal (2003) and the Alps (various times), particularly Matthew Bennett, Michael Chantrey, Nicholas Cox, Kevin Crawford, James Etienne, Ian Fairchild, Becky Goodsell, David Graham, Bryn Hubbard, David Huddart, Wendy Lawson, Geoffrey Manby, Nicholas Midgely, Andrew Moncrieff, Tavi Murray, John Peel, Duncan Quincey and Paul Smith. In addition, numerous other geologists and glaciologists are thanked for leading field excursions in which the author has participated to various other glacier-influenced regions of the world and which figure in this book. Many of these photographs could not have been taken but for the skills of the pilots of chartered helicopters and fixed-wing aircraft, and the boat crews of both large and small research vessels.

In each of the above areas the author's research work has been financed by the UK Natural Environment Research Council, the Swiss Centenarfond, the New Zealand Antarctic Research Programme now known as Antarctica New Zealand, the Alfred Wegener Institute, the Australian Antarctic Research Expeditions, the US Antarctic Program and National Science Foundation, the Leverhulme Trust, the University of Wales (Aberystwyth), The Royal Society and the Transantarctic Association. The work has been assisted immeasurably by the hospitality of the Department of Geography, University of Manchester; the Swiss Federal Institute of Technology in Zürich; the Department of Earth Sciences, Scott Polar Research Institute and St

Edmund's College of the University of Cambridge; the Alfred Wegener Insitute; and the Victoria University of Wellington.

Lastly, I thank my colleagues of the Centre for Glaciology in Aberystwyth, in particular Julian Dowdeswell (now Director of the Scott Polar Research Institute) who invited me to join the Centre, and with whom I have written much about glaciers over the past two decades. Also, Neil Glasser and Bryn Hubbard have each read and commented upon the entire manuscript.

Jürg Alean

Firstly I would like to thank and commemorate the late Eugen Steck who taught me how to observe accurately and patiently as an amateur astronomer, skills which I also found useful later as a glaciologist. Peter Weber instilled in me the interest in, and fascination with, mountain scenery and glacial landscapes. Paul Felber helped me during my struggles when I first tried to study Earth Sciences, and I share with him fond memories of joint undertakings on both tropical and alpine glaciers.

The late Fritz Müller offered me the chance to see and live on the splendid High-Arctic island of Axel Heiberg. Wilfried Haeberli provided guidance as my PhD supervisor and friend, then at the Swiss Federal Institute of Technology. Many other colleagues and friends deserve thanks for their academic and field support. Andreas Kääb, University of Zürich, furnished valuable and up-to-date information on glacier-hazard monitoring and satellite imagery. Most recently, during various educational enterprises involving glaciers and permafrost regions in the Alps, I have enjoyed co-operation and stimulating discussions with Ernst Häne, Walter Hauenstein, Felix Keller, Hans Keller, Max Maisch and Daniel Vonder Mühll.

Most important of all has been my wife's understanding, patience and support, both during our joint global wanderings and when I was working on this book.

We acknowledge the use of modified versions of a number of images and diagrams from other sources. Each is acknowledged alongside the relevant illustration. The drafting of figures was undertaken by Ian Gulley and Anthony Smith, and typing by Carol Parry in the Institute of Geography and Earth Sciences, University of Wales, Aberystwyth. Werner Hartmann and Raimond Reichert (EducETH) helped with various aspects associated with image processing and those kinds of problems that one is guaranteed to have when dealing with computers in general.

Finally we would like to thank Sally Thomas, Earth Science Editor, and her team at Cambridge University Press for advice and guidance during the preparation of this book.

1 Earth, the icy planet

Viewed from space, our Planet Earth displays blue, green, brown and white hues. White indicates not only clouds, but also the **cryosphere**, in other words that part of the Earth dominated by glaciers, ice sheets and sea ice. In this book we focus both on present-day glaciers and ice sheets, and also on the evidence for glacier ice in the distant past. The book draws attention to the relationships between glacier ice and human civilization, and explores what might happen to Earth's ice cover in the future. We hope that the book will convey to the reader not only the beauty and utility of glaciers, but also the nature of glacial landforms and sediments, and help him or her appreciate the vulnerability of this magnificent natural resource and environment.

At the present time, glacier ice covers about 10 per cent of the Earth's land surface. In geological terms, we are living in a glacial era

Distribution of glacierized (glacier-covered) areas of the world (World Glacier Monitoring Service, 1989. World glacier inventory. IAHS (ICSI)-UNEP-UNESCO)

Region	Area (km²)
Africa	10
Antarctica	13 593 310
Asia and Eastern Europe	185 211
Australasia (i.e. New Zealand)	860
Europe (Western)	53 967
Greenland	1 726 400
North America excluding Greenland	276 100
South America	25 908
World total	**15 861 766**

Icebergs on Phantom Lake, Axel Heiberg Island, Canadian Arctic Archipelago.

that began in Antarctica some 35 million years ago. However, the latter stages of this era have included many alternations between periods of full glaciation, when much of the northern hemisphere was covered by ice, and interglacial periods characterized by much less ice, such as at the present day. The most recent full-scale glaciation, when ice covered up to 30 per cent of the land, ended as recently as 10 000 years ago and, if the planet were allowed to follow its natural cycle, a return to such conditions would be expected in a few thousand years' time. However, the big question now is whether the disruption to the Earth's climate system by the burning of fossil fuels is going to be so severe that the enhanced greenhouse effect, already evident in rising global temperatures, will lead to massive reduction of the remaining ice cover. Resolution of this question is a major challenge because different ice masses respond in different ways to climatic change. However, if net melting of glaciers does take place, the consequences will be flooding of low-lying regions of the world.

Figure 1.1 Present-day distribution of glaciers and ice sheets. Note that the areal extent of ice masses becomes increasingly exaggerated towards the poles.

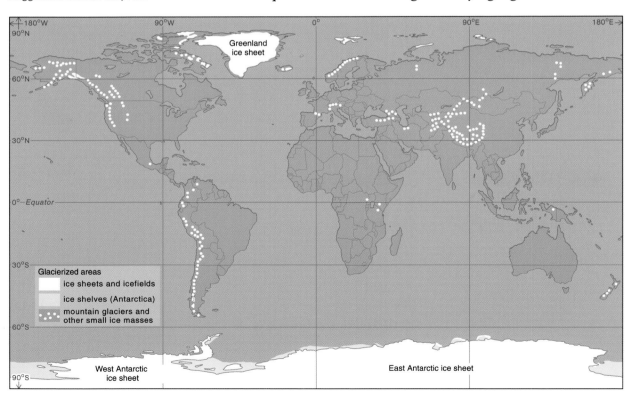

During the last two million years, huge ice sheets advanced across Europe and North America several times. These glacial periods are known as **ice ages**. For example, a major ice sheet developed several times over the Scandinavian highlands, then spread westwards across the North Sea and finally joined a smaller British ice sheet. At its most extensive, ice covered the whole of Great Britain as far south as Bristol and London. From Scandinavia the ice sheet also expanded into northern Germany and Poland. At the same time in the Alps, large glaciers fed by mountain snows carved out deep valleys and spread out over the surrounding lowlands. In North America ice from the Arctic and the western mountain ranges covered almost the whole of Canada and extended south over the prairies of the Mid-West, whilst the areas now occupied by Chicago and New York City vanished under the ice.

Afternoon on Oberaargletscher in the Berner Oberland, Switzerland, looking towards the peak of Oberaarhorn (3638 metres).

In many other parts of the globe similar expansions and recessions of ice took place, notably in the Andes, central Asia and New Zealand. In contrast, the Antarctic and Greenland ice sheets, which survive to this day, appear to have remained largely intact, although there is some debate as to whether the latter disappeared during the last interglacial period, about 130 000 years ago. The relatively small fluctuations of the Antarctic and Greenland ice sheets are related to the fact that glaciers are unstable on reaching deep water, such as at the edge of the continental shelf, and therefore they disintegrate into icebergs.

On a much longer time-scale, the Earth's 4600 million year history has been punctuated by several glacial eras. Continental-scale glaciers occasionally developed at different times on all continents. Even the Sahara and the middle of Australia that are hot deserts today, and the tropical areas of Brazil, bear signs that glaciers once were present,

A glaciated landscape, with the Snowdon range (1085 metres) reflected in Llynau Mymbyr near Capel Curig, North Wales. The mountains last bore small glaciers about 12 000 years ago.

The Norwegian Arctic Archipelago of Svalbard is 60 per cent covered by glaciers. In this spring photograph, heavily crevassed Nansenbreen is flowing rapidly into the sea which is here still frozen.

albeit hundreds of millions of years ago. At such times both the world climate and the relative positions of the moving continents were very different from those of today. Perhaps the most extensive glaciation experienced by the Earth was around six to seven hundred million years ago, and some scientists have argued for an almost completely ice-covered world, known as 'Snowball Earth', at that time.

The importance of glacier ice as a potential water resource cannot be underestimated. Excluding water in the ground, glacier ice represents 80 per cent of the world's fresh water, although 99 per cent of this is locked up in the ice sheets of Antarctica and Greenland, far removed from most human activity. However, in some countries, such as those of central Europe, Scandinavia, the Andes and the Himalaya, mountain glaciers have affected the lives of people for centuries, and not always to their benefit. In these regions, glaciers are an important water resource, capable of supplying water throughout a long, hot, dry period, and considerable amounts of glacial meltwater have been harnessed for the generation of hydro-electricity. In the future, Antarctic icebergs may even provide water for the parched regions of Africa, the Middle East or Australia.

Extensive glaciers survive in the Cordillera Blanca of tropical Peru, but are rapidly receding. Here on Nevado Pirámide (5885 metres), fluted snow on the steep slopes and ice mushrooms on the ridge crests produce avalanches that feed the glacier below.

Yet glaciers can kill. An unwary or careless walker may fall through a snow bridge on a glacier into a hidden crevasse, or be crushed by an ice avalanche. Further from the glaciers themselves, large-scale loss of life has been caused by huge ice avalanches and floods of water that have burst unexpectedly from beneath or in front of a glacier, and valuable pastures, roads and even settlements have been lost under advancing glacier tongues.

Glacial erosion and deposition have an equally important effect on human activities. The slopes of glacially eroded peaks and valleys can be so steep as to be prone to rockfall, but on the other hand give rise to some of the most attractive natural scenery. Glacial deposits over lowland areas provide some of the richest farmland, abundant reserves of sand and gravel for use in the construction industry, as well as local concentrations of valuable minerals.

We know that most mountain glaciers have been shrinking overall since the so-called Little Ice Age of around AD 1750 to AD 1850, although this trend has been interrupted several times by short-lived phases of expansion. In most cases, the long-term recessional trend is generally linked to rising temperatures. Whereas the climate was already warming before industrialization began, the impact of humans through the burning of fossil fuels is leading to unprecedented temperature rises in many parts of the world. As global warming gathers pace, it is predicted that as much as a half of the volume of the planet's mountain glaciers could be lost by AD 2100.

The large ice sheets of Greenland and Antarctica are also sensitive to climatic change, but in ways that we do not fully understand. Rising temperatures have certainly led to recession at the fringes of the ice sheets. On the other hand, the interior of the ice sheets may simultaneously have received increased snowfall, thus triggering growth. Whilst the melting of mountain glaciers is definitely contributing to global sea-level rise, the ice sheets may be counterbalancing this. Nevertheless, we need to take seriously the threat that global warming may have on sea levels, even though no clear relationship has yet been established between rising temperatures and ice-sheet melting.

Admiralty Sound between James Ross and Snow Hill islands off the tip of the Antarctic Peninsula, a conduit through which much of the ice from the break-up of the Larsen Ice Shelf is channelled, creating a hazard to shipping.

Some glaciers are tiny, merely a few hundred metres long, whereas the largest are represented by the great ice sheets of Antarctica and Greenland, and all shapes and sizes exist in between. Naturally enough, the glaciers that have received most attention from glaciologists are those in the mountain regions near the population centres of North America and Europe. The earliest descriptions of glaciers date from the eleventh century in Icelandic literature, but the fact that ice can flow was apparently not recorded until about 500 years later. The first serious scientific studies on glaciers were made in the late eighteenth century, and since then glaciology has been tackled by increasing numbers of scientists trained in geology, geography, physics, chemistry, mathematics and meteorology. Thus, glaciology is a truly inter-disciplinary subject. Nowadays, glaciers are studied on remote Arctic islands, in Antarctica and Greenland and, of course, at more accessible sites such as the Alps, the Rocky Mountains, New Zealand, the Andes and tropical Africa. As a result of all this activity our knowledge of glaciers has improved considerably in the last half century. Now we must use our knowledge to explain how changes to glaciers, and especially their shrinking under warmer climates, will have an impact on humans throughout the world.

Tabular iceberg in the northwest Weddell Sea. An ocean current carries icebergs from the southern limits of the sea up the east coast of the Antarctic Peninsula, before spewing them out into the Southern Ocean.

Nuptse (7861 metres) forms the south west end of the ridge that begins with Mount Everest and encircles the great glacier amphitheatre known as the Western Cwm. This view of the peak is from the western lateral moraine of Khumbu Glacier, which it supplies through avalanching.

2 The glacier family

The largest glacier in the European Alps, the Grosser Aletschgletscher, from the air. The glacier descends for 23 kilometres from peaks of over 4000 metres, and supplies meltwater for a major electrical power station owned by the Swiss railway system.

Glaciers are usually classified according to their shape and their relationship with the surrounding and underlying topography, but some are described on the basis of the temperature distribution within the ice. However, we need to bear in mind that these distinctions are not strict and that transitions exist between all these types of glacier.

Glaciers classified according to topographic setting

Ice sheets and ice caps

The largest glaciers are the **ice sheets** of Antarctica and Greenland. Covering a continent twice the size of Australia, the Antarctic Ice Sheet contains 91 per cent of the world's freshwater ice. This ice sheet is more than 4000 metres thick in some places, inundating entire mountain ranges. In much of West Antarctica, ice rests on bedrock that is many hundreds of metres below present sea level. Consequently, if the West Antarctic Ice Sheet were to melt, a sea with numerous island archipelagos would emerge. In contrast, the ice sheet over East Antarctica rests on ground that is mainly above sea level, but drains radially via a number of valleys, some of which are

Figure 2.1 Cross-section through the East and West Antarctic ice sheets, illustrating the irregular nature of the bedrock and the ice thickness, as well as floating ice shelves. (Adapted from D. J. Drewry, 1983, Glaciological and Geophysical Folio, Scott Polar Research Institute).

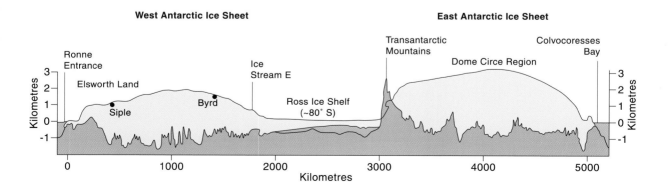

Discharge from the polar ice sheets is commonly via outlet glaciers. Here, the upper Shackleton Glacier, flowing from left to lower right, together with its tributaries, is leaving the Polar Plateau in Antarctica, and will eventually drop down to sea level and form part of the Ross Ice Shelf.

below sea level. Apart from the Transantarctic Mountains and the mountainous backbone of the Antarctic Peninsula, which are high enough in places to project above the level of the ice sheet, rock outcrops are few and far between. Isolated rocky mountains surrounded by ice are known as **nunataks**, a word from the Inuit language. Although much smaller, the Greenland Ice Sheet with eight per cent of the world's freshwater ice nevertheless covers an area the size of Mexico or ten times that of the British Isles. The Inlandis (Inland Ice), as it is known, fills a huge basin that is rimmed by ranges of mountains to a depth of over 3000 metres. Ice overflows and breaches this rim in many places, discharging into the sea and producing icebergs.

Ice caps are geometrically similar to, but smaller than, ice sheets. An ice cap is defined as covering an area of less than 50 000 square kilometres, but there are many that cover only a few square kilometres. Like ice sheets, they inundate the underlying topography, and their smooth surface belies the irregular nature of the bed. Ice caps are common in polar and sub-polar regions, but are rare in regions of alpine topography. They tend to develop on high plateaux from which

A small ice cap on the northeast side of James Ross Island, Antarctic Peninsula region, forms a white dome-shaped feature, capping the dark grey volcanic rocks below.

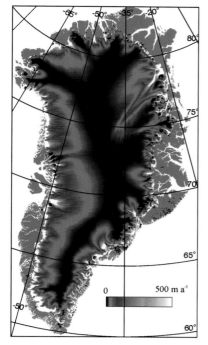

Satellite-derived image to illustrate the Greenland Ice Sheet. Satellite data can be processed in many different ways. In this case, the flow pattern of the ice sheet has been derived, the colours representing different velocities in metres per year, as the scale indicates. Black areas show slow speeds and define the several drainage basins that make up the ice sheet. White areas illustrate the high speeds achieved by outlet glaciers. The mauve area surrounding the ice sheet is ice-free land, except for small local glaciers. (Image provided by Jonathan Bamber.)

discharge is inhibited by the low gradients, except where ice spills over the plateau edge. Amongst the largest are Austfonna and Vestfonna on Nordaustlandet in the High-Arctic archipelago of Svalbard, covering an area the size of Wales or Connecticut. In temperate regions, the largest ice caps are Vatnajökull in Iceland and the North and South Patagonian ice caps that straddle the southern Andes.

Both ice sheets and ice caps discharge in a variety of ways; for example, through valleys on land, by breaking up as the ice flows off the cliff (an **icefall**) bounding the plateau, or directly into the sea. Ice sheets and ice caps that flow into the sea have a unique feature – **ice streams**. These are zones of much faster flow that have well-defined boundaries with slow-moving ice; they behave like separate glaciers and commonly have a heavily crevassed surface, whereas the bordering ice is far less disturbed. Their margins are zones of very intense shearing, reflected in the development of deep, closely spaced crevasses. Some ice streams in Antarctica extend into the sea as unconstrained floating **glacier tongues** which periodically break off as huge icebergs.

Ice shelves

In Antarctica there is a net accumulation of snow and ice close to sea level, which encourages the formation of **ice shelves** – slabs of

The highland icefield of Lomonosovfonna, in northeastern Spitsbergen, forms an elongated spine of ice through which mountains of around 1500 metres in height project. From this icefield, glaciers descend in different directions, most of them reaching the sea.

glacier ice that float on the sea. Ice shelves typically range in thickness from over two kilometres in their inner parts to 200 metres at their terminus where they produce icebergs, a process known as **calving**. Those glaciers that discharge into the sea from higher ground become detached from the bed and float, spreading out to cover large bays, as happens in the Ross and Weddell seas in Antarctica. Some ice shelves cover vast areas; the Ross Ice Shelf, for example, measures about 850 by 800 kilometres and covers over half a million square kilometres, equivalent to the size of France.

Periodically, large parts of these shelves break off as smooth, flat-topped, or 'tabular', icebergs, sometimes measuring more than 100 kilometres across. Some ice shelves remain stable over hundreds of years, but others can disintegrate rapidly. As we shall see in Chapter 8, several ice shelves in the Antarctic Peninsula have largely disappeared in the past 30 years. Except for the Ward Hunt Ice Shelf on Ellesmere Island in the Canadian Arctic, and another in the Russian Arctic archipelago of Severnaya Zemlya, which are tiny in comparison, ice shelves are peculiar to the Antarctic.

The smallest glaciers are only a few hundred metres across, such as this tiny cirque glacier (Teton Glacier) in Grand Tetons National Park, Wyoming, USA. Despite its small size, the glacier has produce a sizable terminal moraine (the bare ground in front), although it may be ice-cored and the final landform may end up being just a few metres high.

Mountain glaciers

Often as extensive as ice caps, **highland icefields** are semi-continuous sheets of glacier ice occupying many square kilometres, burying many of the features of the underlying landscape. They are most common in polar and sub-polar regions, such as Spitsbergen, the Queen Elizabeth Islands in the Canadian Arctic, southeast Alaska and the Yukon, Patagonia and parts of the Antarctic Peninsula. Smaller ones are found in other high mountain areas in more temperate latitudes. The highest mountains project above the ice as nunataks, between which the ice surface is undulating. This ice surface roughly mirrors the topography beneath. **Valley glaciers** commonly flow out in several directions from those icefields that are located in temperate regions.

High-altitude ice may flow downwards from ice caps, highland icefields, ice sheets and amphitheatres known as **cirques**, as **outlet glaciers**. These tongues of ice, typically tens of kilometres long, flow downhill into regions well below the snow line, even occasionally into temperate rain forest, as in Alaska, New Zealand and Chilean Patagonia. Some of the most spectacular examples descend into the Pacific Ocean from the highland icefields of southern Alaska and the Yukon. Bering Glacier and Hubbard Glacier (191 kilometres and 150 kilometres long respectively) are the longest in the Americas. In

Cold glaciers entering water bodies typically have floating glacier tongues. In the Amery Oasis of East Antarctica we see the local mountain glacier, Battye Glacier, descending into the mostly frozen Radok Lake. The featureless white background is the Amery Ice Shelf.

contrast, mainland Europe's longest valley glacier, the Grosser Aletschgletscher, is a mere 23 kilometres long, but impressive enough to those who walk across it or view it from neighbouring summits.

In high latitudes, many valley glaciers enter the sea, where they either remain grounded or float, and are then known as **tidewater glaciers**. Ice velocities usually increase as the glacier enters water, producing exceedingly crevassed glacier termini. The icebergs that are produced by calving are normally small and irregular in shape when they originate from grounded tidewater glaciers, but there are recorded instances of large tabular bergs from floating glaciers. Impressive tidewater glaciers are found in the fjords of Spitsbergen, Greenland, Arctic Canada, Alaska, Patagonia, South Georgia and the Antarctic Peninsula.

Cirque glaciers such as Skoltbre on the peak of Oksskolten (1916 metres) in Nordland, Norway, have formed some of the most attractive glacial landforms in places like the British highlands or the Rockies. Below the cirque is the heavily crevassed valley glacier, Austre Okstindbre, which descends from a small ice cap.

Fast-flowing glaciers that reach the sea discharge large numbers of icebergs. Kronebreen in northwest Spitsbergen is one of the most prolific in the Svalbard archipelago, and bergs normally litter the beaches of the scenic Kongsfjorden.

Where mountain valleys open into larger valleys or on to plains, valley glaciers spread out into wide lobes called **piedmont glaciers**. The stunning Malaspina Glacier in south-east Alaska, measuring 70 kilometres across, is the best known.

In some mountainous regions ice accumulates on slopes that seem to be almost too steep to hold any snow. Smooth ice slopes are known as **ice aprons**, whereas unstable bulging crevassed ice masses are **hanging glaciers**. Both types are generally small, being only a few hundred metres across at most. Ice frequently breaks off hanging glaciers, forming avalanches that are a particular hazard to anyone walking or climbing beneath. Large ice avalanches have sometimes wiped out whole villages in the Alps (as described in Chapter 13). Ice debris from such avalanches, and also from the snouts of tributary valley or cirque glaciers that overhang the main valley, may build up sufficiently to produce a **rejuvenated** or **regenerated glacier**.

Warm and cold glaciers

A different way of classifying glaciers is according to their temperature distribution, in which respect there are three basic types. Firstly, **temperate** or **warm** glaciers are those in which the ice is at the

Opposite. The Stauning Alper of East Greenland are classic alpine peaks of granite, separated from the main ice sheet, but heavily eroded by valley glaciers. The highest peak is Dansketind (2930 metres), as shown here, from which a valley glacier descends coastwards for several kilometres.

Glaciers that flow through narrow valleys and then reach an open plain spread out as broad lobes known as piedmont glaciers. These striking examples are located near Surprise Fjord in the southern part of Axel Heiberg Island, Canadian Arctic.

melting point throughout, although a thin surface layer cools below zero degrees Celsius in winter. Meltwater is abundant in summer and generally a small amount continues to be discharged by the glacier even in winter. Meltwater normally emerges through a tunnel called a **glacier portal** in the middle of the glacier snout. However, beneath or within the glacier there usually is a well-developed drainage network, as well as a number of water pockets. Warm glaciers are characteristic of most mountain regions outside the Arctic and Antarctic.

Secondly, in the much colder Polar Regions, where the mean annual air temperature is at least several degrees below zero, much of the ice is well below the melting point. Where the bulk of the ice is below the melting point, the glacier is termed **cold**. In the upper 12 metres or so the ice temperature fluctuates according to the season, but below that depth it is similar to the mean annual air temperature. Further towards the glacier bed, heat flow from the bedrock warms the ice, perhaps to the melting point. Warming of basal ice is partly determined by pressure; if thick enough the ice will

Glacier ice can cling to surprisingly steep mountainsides where it builds up to form avalanche-prone hanging glaciers as here on the northeastern peak of the long ridge of Ombigaichan (over 6000 metres), Khumbu Himal, Nepal.

melt at the bed and cause basal sliding. This is true even in the heart of Antarctica where subglacial lakes have been located beneath thousands of metres of ice, and where as much as one-third of the ice sheet is wet at its base. Cold glaciers in the Arctic produce abundant meltwater during the brief summer, but their drainage pattern is

The thermal regime of glaciers influences the manner in which they flow and control the distribution of meltwater. Determining the thermal characteristics of a glacier, such as the White Glacier of Axel Heiberg in the Canadian Arctic shown here, involves the use of a hot water drilling system which penetrates to the bed and allows scientists to insert sensors to automatically record temperature over several years.

different from that of temperate glaciers: stream channels only develop on the surface or close to the glacier margins, as the streams which cut downwards into the ice cannot survive for long without freezing up.

Thirdly, the term **polythermal** is commonly applied to glaciers that have both warm and cold parts. Typically, the snout and margins of a polythermal glacier are frozen to the bed, but the thicker upper part may be wet at the base since geothermal heat has greater influence than air temperature beneath thicker ice. Glaciers of this type are abundant in the High-Arctic, notably in the Canadian Arctic Archipelago and Svalbard, and also in sub-Antarctic regions. However, some polythermal glaciers occur on very high mountains in temperate latitudes as well. One example is Grenzgletscher, fed by peaks in excess of 4000 metres, situated near the famous resort of Zermatt in Switzerland.

The thermal characteristics of a glacier have a strong bearing on how the underlying landscape is modified. In regions once covered by ice, meltwater channels and the distribution of stream deposits reflect the type of glacier that occupied the area. Temperate glaciers and those parts of cold glaciers that slide on their bed erode the terrain strongly. Cold glaciers that are frozen to the bed are relatively

Many glaciers of polar regions contain at least some ice which is colder than the pressure melting point, particularly near the surface and along the rim. They are then called polythermal. Glaciers whose ice is entirely 'cold', i.e. temperatures below the melting point throughout, are common only in Antarctica.

passive. Although they can erode the bed, they do so by subglacial deformation and incorporation of debris into the basal part of the glacier. Commonly, cold glaciers protect the land from weathering and other kinds of erosion, such as by wind or streams. Regardless of the topographic and thermal classifications used, the types described above are merely convenient designations within a continuous spectrum of types of ice masses, and many glaciers are a combination of types.

3 Birth, growth and decay of glaciers

High elevations in the tropical Andes receive large quantities of snow during the rainy season in summer. This peak of Nevado Parón (5600 metres) bears the characteristic mushroom-like growths of snow and ice, as well as ice cliffs that feed into the glacier below.

Glaciers are sometimes called 'rivers of ice'. However, this is misleading since glaciers do not normally form from rainfall, but by the transformation of snow to ice. To initiate a glacier, winter snowfall needs to be great enough for some of the snow to last throughout the following summer. This process is then repeated for several years. Finally, under the pressure of its own weight the snow turns into ice. If the ice is thick enough, it flows under the influence of gravity. This transformation of snow to ice is often a long and complex process, since both the nature of the transformation and the time involved depend on temperature and the depth of further, overlying snow. The transformation is most rapid in temperate regions, such as the Alps and the Western Cordillera of North America, where ice can form from snow within five to ten years. In contrast, the transformation in high polar latitudes or at high elevations may take hundreds of years.

From snowflake to glacier ice

Although snow crystals tend to have a hexagonal structure, with characteristic six-sided symmetry, snow falls in myriad forms. Snowflakes may come as delicate, feathery crystals a centimetre or

Figure 3.1 Transformation of snow to glacier ice crystals. The times indicated are typical for a temperate glacier.

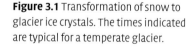

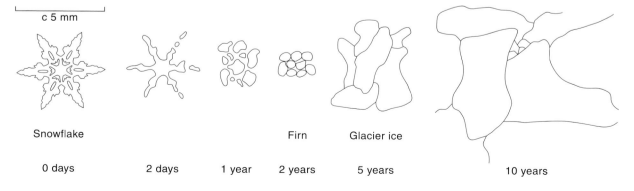

| Snowflake | | Firn | Glacier ice | |
| 0 days | 2 days | 1 year | 2 years | 5 years | 10 years |

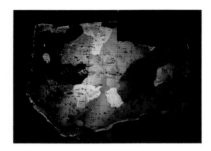

The transformation of snow to ice results in interlocking crystals and entrapment of bubbles, as seen in this thin section of ice from the north Norwegian glacier; Charles Rabots Bre. The photograph was taken through crossed polaroid plates which brings out the crystal structure. The straight lines represent a centimetre grid.

In slow moving or stagnant temperate glaciers, ice crystals can attain large sizes, as this example from Columbia Glacier, Alaska demonstrates.

so across, or as relatively hard grains that have the feel of sand. They have their most intricate and varied forms when they fall close to freezing point, and can form a very light snow layer 20 times less dense than water. Snow like this is fluffy – the powder snow much beloved of skiers.

A glacier can form where the annual accumulation of snow exceeds the amount of melting or evaporation each year. In high mountain regions, the net accumulation of snow and its conversion to ice is related to temperature, which in turn is dependent on altitude. The amount of snow is also significant: wind leads to greater accumulation on the lee (downwind) side of a mountain than on the windward side. Also, understandably, the shady sides retain more snow.

Even meltwater has a bearing on the formation of a glacier. In temperate areas, where meltwater is abundant, snow is converted to ice in a number of stages. Firstly, the fragile crystals break on settling, as they are compressed by the weight of additional snow on top, or they break down if they become wet. Gradually, snowflakes change to grains, which become rounded and granular like coarse sugar. As the snow becomes compressed it becomes harder and denser. At first the air spaces between the grains are connected, but now the snow is called **firn**, a term derived from the German, meaning 'old snow'; this is an intermediate stage in the transformation to ice. The firn stage is generally reached after one year, when the snow's density has reached half that of water.

As these changes proceed, the roundish grains of firn begin to recrystallize and larger crystals of ice begin to form at the expense of their smaller neighbours. Air is now only present as bubbles trapped inside the growing crystals. These changes are aided by the flow of the glacier, as ice crystals deform readily under gravity-induced stress. In a flowing glacier the form of the crystals is in constant change. If the ice is deforming rapidly the crystals may not grow very large because they never remain stable long enough. However, by the time the ice reaches the glacier terminus, where flow rates and therefore stresses are lower, crystals may grow to several centimetres across. In stagnant ice they may grow even larger, up to 25 centimetres in length. They then have a very complex shape. By this time ice has reached a density of around 90 per cent that of water.

In glaciers where there is no surface melting at all, as at high altitudes in Greenland or Antarctica, the change from snow to ice may take several hundred years, rather than a few years. The transformation here is determined by three factors: the movement of crystals relative to each other, the effect of increasing compaction under the accumulating snow and internal deformation.

Observers are often struck by the blueness of some glacier ice, which is especially noticeable on cloudy days. This is because water molecules preferentially absorb all colours of the light spectrum except blue.

Climbers ascending Weissmies (4023 metres) in the Swiss Alps are dwarfed by the crevasse they have just crossed. The crevasse wall clearly shows the accumulation layering, the most prominent of which bound a year's accumulation of snow. The yellowish layers mark deposits of Saharan dust which is commonly transported by wind over central Europe.

Profit and loss

The changes of snow to ice and movement down-slope are manifested in terms of the balance between profit (**accumulation**) and loss (**ablation**). Glaciologists refer to this concept as the **mass balance** or **mass budget**. A bank account is a good analogue of mass balance. If we add more (snow and ice) to the account than we remove (as meltwater or by calving), then our account builds up credit. If accumulation in a glacier exceeds ablation, the glacier has a positive mass balance. Conversely, if ablation exceeds accumulation, the glacier has a negative mass balance equivalent to a bank account in deficit.

We can observe the changes from profit to loss by walking down a typical alpine or Arctic valley glacier in late summer. In the upper

Snow accumulation layers in a glacier cliff bordering Granite Harbour, western Ross Sea, Antarctica. The base of the basin-like form defines a break in deposition and is referred to as an unconformity.

Blowing snow on Nevado Huascarán (6768 metres), Cordillera Blanca, Peru. Accumulation of snow on tropical mountains normally takes place during the rainy season, rather than during a (non-existent) winter.

reaches, where snow accumulation exceeds ablation, we are in the **accumulation area**. Its highest part may consist of a dry snow zone where there is never any surface melting. However, only polar or very high altitude glaciers have such a zone. Passing downwards we enter a zone where some melting occurs. Water percolates through the surface and refreezes below, creating ice layers, ice lenses, or pipe-like structures known as ice glands. Here, the glacier is still gaining mass.

Continuing further downwards we pass into the **wet snow zone**, where most, if not all, the winter snowfall gains have been lost. All the snow here is raised to melting temperature and becomes saturated with water. On Arctic glaciers large areas of slushy snow are formed in this area, much of which refreezes in winter to form **superimposed ice**, which can add to the glacier's mass. The boundary between the superimposed ice and wet snow zones is known as the **firn line**. On glaciers in temperate regions, however, this superimposed ice zone is generally rather narrow. In contrast, in many parts of Antarctica, the accumulation area extends down to sea level, and ablation is mainly through the calving of icebergs.

Figure 3.2 Long profile through a glacier, illustrating the accumulation and ablation areas. The associated flow paths of material as it becomes buried and emerges later at the surface are also shown.

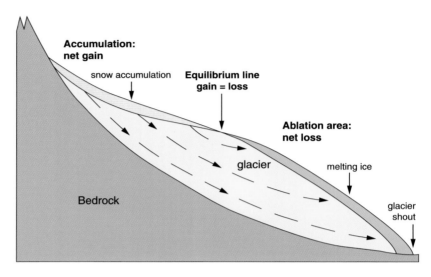

The lower limit of superimposed ice represents the most important boundary in mass-balance terms. It is referred to as the **equilibrium line**. In temperate regions this line coincides approximately with the firn line. Here profit equals loss. Below the equilibrium line, in what is known as the **ablation area**, there is a net loss of ice. All the previous winter's snowfall has melted, as well as an increasing amount of ice. In the lowest part of a typical Alpine glacier the melting of the ice vertically may exceed 10 metres a year, producing increasing mass deficit as we move towards the glacier terminus. However, this does not necessarily mean that the glacier recedes, since the lost ice may be replenished by ice flowing from above.

In early summer large areas of melting snow in the ablation zone become saturated with water, especially if the glacier surface is flat. Such **snow swamps** are common on glaciers in the Polar Regions because cold ice (ice below melting point) prevents meltwater from draining freely.

As summer progresses, ice crystals in the ablation zone melt along their boundaries in sunny or dry weather, so that the glacier surface acquires a knobbly texture that is easy to walk on, even without crampons. In contrast, in rainy weather ice melts evenly and becomes much more slippery. Sunny weather also enhances the surface undulations, especially if there are uneven concentrations of debris,

The effect of the Sun's radiation on bare ice is to cause fine debris and stones to sink faster into the ice, producing cryoconite holes. These may remain filled with meltwater in the evening, then freeze over at night, trapping air bubbles which act as miniature wide angle lenses.

The accumulation of snow in the heart of Antarctica amounts to only a few centimetres per year, and much of that is reworked by wind. Here fierce winds at the edge of the Polar Plateau on Roberts Massif have carved the snow into massive flower-like structures.

Below. Ice avalanches in high mountain terrain contribute to the build-up of ice in the accumulation area, as here on Nevado Chacraraju (6172 metres), Cordillera Blanca, Peru.

At the end of the summer season, alpine glaciers have a clear demarcation between the accumulation area and the ablation area known as the equilibrium line, as here on Vadret Pers, Engadin, Switzerland. The bluish grey ice in the foreground would not be sustainable at this lower level were it not for the ability of ice to flow by sliding and internal deformation from higher levels.

Opposite. Glaciers respond to a negative mass balance either by recession of the snout or down-wasting. Here Mueller Glacier in the Southern Alps of New Zealand is down-wasting into an incipient lake. Its Little Ice Age lateral moraines are in the background and behind them is the country's highest peak, Mount Cook, or Aoraki in Maouri, meaning the 'Cloud Piercer' (3754 metres).

and can produce pinnacles called **ice ships.** These are often a metre high, but examples several metres in height are found where solar radiation is very strong, in particular at low geographical latitudes, such as in the tropical Andes and the Himalaya.

Response of glacier terminus to mass-balance changes

Changes in the overall or net mass balance of a glacier are reflected in changes to the lower reaches of the glacier or **tongue**. In particular, the response of the ice margin or **snout** at the terminus helps us to understand the state of health of a glacier. A 'healthy' glacier is one in which as much ice is formed in the accumulation area as is lost in the ablation area, in which case the snout remains in the same

Rapid recession of glaciers often leads to masses of ice becoming detached from the snout and producing tunnels and caves, as here at Haut Glacier d'Arolla, Valais, Switzerland.

Glacier snouts that are completely covered by debris in the Himalaya generally lose mass by down-wasting most where the debris cover is thinnest, that is closer to the equilibrium line. As a result glaciers can gain a reverse gradient, that sometimes allows large lakes to develop. This view is of the debris-mantled tongue of Khumbu Glacier in the Everest region of Nepal.

position. If gains exceed losses, the glacier will ultimately advance, but it may take many decades for the trend to be reversed. An 'unhealthy' glacier loses more mass than it gains, and its ice margin will recede or waste away through surface lowering.

The Southern Alps of New Zealand provide a remarkable contrast in the response of glaciers to changes in mass balance. On the maritime west flank of the Mount Cook range, precipitation over the Fox and Franz Josef glaciers reaches the phenomenal value of 10–15 metres a year of water equivalent, much of it falling as snow at high altitude. The glaciers descend into rain forest where it hardly ever snows, at velocities reaching an annual average of 700 metres a year. In periods of heavy rain, when the bed is lubricated, these glaciers may even attain speeds equivalent to 2500 metres a year. The accompanying mass-balance differences are reflected in a very rapid response of the glacier snouts, with advance and recession rates typically of a half to one kilometre a year. Rapid advances through the last two decades of the twentieth century have been superseded by equally rapid recessions since 1999.

Small Antarctic glaciers, separate from the ice sheet, show steep vertical cliffs if their snouts are stationary or advancing. This cliff, with associated avalanche debris, belongs to the Rhone Glacier (named after its Swiss counterpart) in the Dry Valleys region.

A few kilometres across the Mount Cook divide, the glaciers are much less dynamic. Here the precipitation drops off rapidly, so that near the snout of the Tasman Glacier it is only 0.4 metres a year. Although this glacier is the biggest in New Zealand, it has a maximum velocity of only 250 metres a year. Also its snout does not advance rapidly forwards or backwards like those of Fox or Franz Josef glaciers. Rather, it has thinned over several decades along its entire ablation area or glacier tongue. Only in recent years has the glacier begun to show ice-frontal recession; it is doing so by expansion of a terminal lake. This behaviour highlights an aspect of glacier fluctuations that is discussed more fully in Chapter 4, that the snouts of short, steep glaciers are more susceptible to fluctuations than long, gently graded ones.

Glaciers terminating in the sea (tidewater) are subject to more rapid fluctuations than those on land. They are characterized by vertical calving cliffs, grounded on the sea bottom, as in this view of Hubbard Glacier, Alaska. Unlike most glaciers in Alaska the Hubbard has been advancing strongly for the last century.

In the European Alps, and many other areas, it is sometimes possible to observe both advancing and retreating glaciers in adjacent valleys under the same climate. This behaviour was last evident in the Alps in the 1970s and early 1980s, but in Scandinavia we see this pattern emerging in the late 1990s and early 2000s. How can this apparent paradox be reconciled with mass balance? Small glaciers tend to respond within a few years to changes in mass balance, whereas the larger Alpine valley glaciers like the Grosser Aletschgletscher may take half a century or more. For large glaciers, then, short-term variations in mass balance will not be reflected in changes in the snout position; they will only respond to longer-term changes in climate.

Generally, climatic changes over a period of several decades will

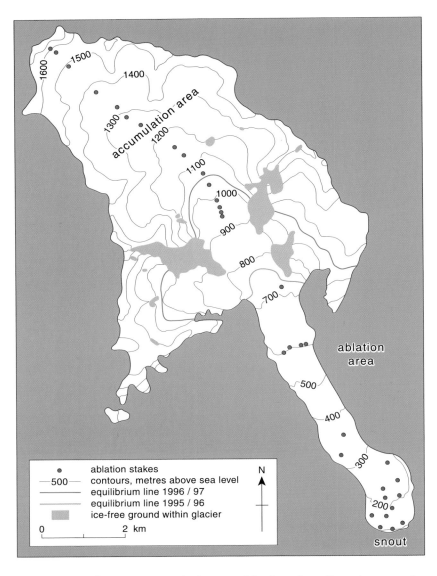

counteract short-term variations. This is why climatic warming
since the end of the nineteenth century has led to almost all the glaciers in North America and Europe receding considerably, despite the
temporary readvances of some of them. For some glaciers, warming
may be accompanied by increased precipitation. If this falls as snow,
then the glacier may acquire a positive mass balance, as in parts of
Norway at the end of the twentieth century.

In contrast to valley glaciers, such as those referred to above, polar ice sheets and ice caps respond much more slowly to climatic changes. Indeed, the influence of warming that accompanied the end of the last ice age around 10 000 years ago has still not fully penetrated the main ice sheets. In fact, the ice sheets may never reach equilibrium. On many parts of the East Antarctic Ice Sheet precipitation rates are only a few centimetres a year. Ablation is predominantly by calving into the sea, rather than by melting. As the several ice drainage basins that make up the East Antarctic Ice Sheet each cover up to a million square kilometres, changes in mass balance need hundreds of years to be transmitted to movements in the position of the ice margin. Furthermore, changes of the ice margin may

Mass-balance studies are labour-intensive operations that involve drilling holes with hot-water drills as here on Arteson Raju, Cordillera Blanca, Peru. Wooden, metal or plastic stakes are inserted into the holes, and the distance from the top to the ice or snow surface is measured at the beginning and end of the ablation season.

This group of smart wooden buildings is the Tarfala Research Station in the Kebnekaise region of northern Sweden, operated by Stockholm University. It was here that the first mass balance programme was initiated immediately after World War Two, and continues to the present day. It provides a unique record of glacier response to climatic change over more than half a century (see graph in Figure 3.4).

The glacier selected for mass-balance investigations was Storglaciären, the surface of which is shown here after a mid-summer snowfall. The peak in the background, Kebnekaise (2111 metres), the highest in Sweden, has a thin ice cap.

be controlled more by oceanographic factors, such as the temperature or the level of the sea, than by changes in mass balance.

Measurements of mass balance

Measuring the mass balance of a glacier is a labour-intensive operation. Firstly, wooden or aluminium stakes must be drilled into the ice at frequent intervals along a longitudinal profile and on a series of transverse profiles as the accompanying map (Figure 3.3) demonstrates. Typically the stakes are five or six metres in length and drilling is undertaken with a hot-water or steam drill, operating on propane gas. In the accumulation area, stakes are left projecting from the glacier surface sufficiently to prevent burial by the winter snow. They must remain visible until the next spring. Deep snow pits have to be dug and samples taken in order to determine how snow density varies with depth.

In the ablation area, the stakes are inserted flush with the ice surface, as commonly several metres of melting will occur during the summer. If the stakes threaten to fall out they have to be redrilled. All the stakes then have to be surveyed to record their positions. Resurveying them later allows us to calculate ice velocities as the glacier carries them down-valley. The elevations of the ice and snow surfaces are measured at the beginning and end of the ablation season, and the changes of mass calculated in terms of water-equivalent. Mass-balance contours in metres can then be drawn on the maps, and total volume changes determined. This procedure must be repeated year-after-year, as the data are only of value when a decade or more has been covered. As funding agencies generally prefer to support short-term projects, long-term monitoring is rarely given the support it requires.

It is, of course, logistically impracticable to undertake this kind of investigation on a large ice mass. For the Greenland and Antarctic Ice Sheets satellites are becoming precise enough to record changes in surface elevation over a number of decades. However, there is still a need to validate any results by undertaking over-ice traverses and

Figure 3.4 The world's longest mass-balance record was obtained from Storglaciären in northern Sweden, and covers the period 1945 to the present. The zero mass-balance line indicates that the glacier in that year has neither gained nor lost mass. Bars above this line indicate net gains in mass (positive balance) and bars below net losses (negative balance). Data supplied by Peter Jansson, Stockholm University.

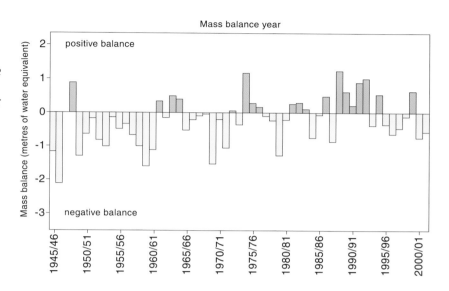

measuring snow density and accumulation rates on the ground. But even this may become a thing of the past as satellites can now measure gamma-ray emission that indirectly records ice and snow densities.

The UNESCO-supported World Glacier Monitoring Service, based in Zürich, assembles mass-balance data from around the world. This data-set is crucial for predicting glacier response to climatic change, and the implications for global sea-level changes and water resources. In the latest compilation, summary data are available for 88 glaciers in 16 countries. This is a very small sample with which to assess global trends. Unfortunately, several countries have wound down their mass-balance programmes in recent years because of reduced government funding. There is a strong bias towards the European Alps and Scandinavia, and data are urgently needed from other glacier-covered regions of the world.

4 Fluctuating glaciers

Glaciers are one of the Earth's most sensitive indicators of climatic change. However, evaluating the significance of ice-marginal fluctuations is far from straightforward, as glaciers respond to climatic warming and cooling on different time scales. At the beginning of the new millennium, the majority of the Earth's mountain glaciers are undergoing recession, although there are some notable exceptions.

Along with the recession of glaciers comes the transfer of water stored on land to the sea, and with it the fear of rising global sea levels. Hence it is vitally important that we acquire data of the past and present response of glaciers to climatic change, in order to help us predict the future. As a step towards this goal, the World Glacier Monitoring Service assembles data on ice-marginal fluctuations for several hundred glaciers around the world.

Historical records, dating back to the eighteenth and nineteenth centuries, have been the prime source of information about glacier fluctuations, but there are few such records outside the Alps and Norway. In Europe this period was typified by the strongest advance of glaciers since the last ice age and became known as the Little Ice Age. It coincided with low average temperatures and widespread crop failure. In these and other regions dating of old ridges of glacial debris (**moraines**) by radiocarbon techniques has enabled scientists to determine glacier fluctuations on a time-scale of thousands of years. More recently, satellite imagery has become an essential tool for monitoring glaciers worldwide, especially in Greenland and Antarctica. The first spacecraft used for this purpose was the Earth Resources Technology Satellite (ERTS-1 or Landsat-1), launched in July 1972. It was followed by many successful satellites that are now used for monitoring ice masses.

The decline of glaciers is usually manifested by recession of the ice margin, but this is not the only indicator. Some glaciers shrink

Gulkana Glacier in Alaska is one of the best studied in North America. This land-based glacier has thinned and receded dramatically in the last few decades, and now has the relatively flat snout that is typical of receding glaciers. Prominent moraines at the surface grow as more englacial debris is exposed by ablation.

Triftgletscher in the Berner Oberland of Switzerland until recently filled the entire basin seen here in the foreground. Thinning of the tongue during the 1990s accelerated, and by 2001 a lake had started to form in front of it (left-hand photograph). The ice then became buoyant and rapid break-up of the snout was underway by 2003 (right-hand photograph).

mainly by down-wasting of the ice surface, especially if they are debris covered.

One complication in assessing whether glaciers in a particular region are advancing is that adjacent glaciers can behave quite differently from each other. For example, on Axel Heiberg Island in the Canadian High Arctic, there are two large valley glaciers, whose snouts are actually in contact with each other. Since observations began in 1959, the 14.5-kilometre-long White Glacier has receded consistently at a rate of a few metres a year, behaviour that is reflected in the smooth rounded profile of its snout. In contrast, the 35-kilometre-long Thompson Glacier, which is a major outlet from the island's largest ice cap, has been advancing dramatically. Its

Above. Thompson Glacier on Axel Heiberg Island has shown sustained advance since at least the 1960s when scientific investigations began in the area. The 30-metre-high cliff is characteristic of large advancing polythermal glaciers. In this case it is not only overriding a 'meadow' of well-established purple saxifrage, but is also pushing up a large pile of river gravel.

An oblique aerial view of the contact zone between the White Glacier (left) which is slowly receding, and Thompson Glacier (right), which is advancing (Axel Heiberg Island, Canadian Arctic Archipelago). Note the difference in the nature of the glacier frontal zones.

spectacular, 30-metre-high, vertical cliff has been moving forwards at an average rate of 15 metres a year, sometimes collapsing and overriding its own debris. Plant life on the tundra that has taken several thousand years to become established is now being destroyed as the ice advances across it. In addition, Thompson Glacier is pushing a huge pile of river gravel forwards, like an enormous bulldozer.

Fluctuations in the European Alps

Small ridge of boulders (a push moraine), bulldozed by Begsetbreen in western Norway during a recent advance, but now separated from the glacier as it begins to recede.

Nowhere else on Earth have glacier fluctuations been recorded in greater detail than in the European Alps. Farmers have carefully recorded the advance and recession of glacier tongues, as meadows disappearing under advancing glaciers represented a serious loss to

Advancing glacier, Brikdalsbreen, western Norway, flowing from the Josterdalsbreen ice cap, is pushing loose debris in front of it and ploughing into recently established birch woodland. The steep, heavily crevassed front is typical of such glaciers.

the local population. Over the centuries pastures above the timber line have been used for grazing cattle and sheep in summer, while many of the mountain passes have served as important merchant routes, so the memory of their loss has been passed down in legends and official records for hundreds of years. A frequently cited legend is of diabolical beings (that is the glaciers) eating up the grasslands, as around the beautiful Swiss village of Tiefenmatten, which was situated north of the Matterhorn, but which was buried by the advancing Z'Muttgletscher.

During the last few hundred years, tourists, especially painters and mountaineers, have documented glaciers in the Alps. The Untere Grindelwaldgletscher in the Bernese Oberland has been painted so accurately and frequently that the history of its advance and recession has been reconstructed with considerable precision as far back as 1600.

In 1890, systematic glacier measurements were initiated in Switzerland, and they are continued to this day. Forest service workers and glaciologists who took on this considerable task of fieldwork have become witnesses to the dramatic recession of many Alpine glaciers. The Grosser Aletschgletscher, the largest glacier in the Alps, lost 2.2 kilometres within the twentieth century. This means that the glacier retreated, on average, 22 metres per year between 1900 and 2000.

In front of other, more easily accessible, glaciers in Switzerland, markers have been inserted to show visitors where the terminus was situated in former times. A particularly rewarding example is the walk from the railway station at Morteratsch (near the well-known resort of Sankt Moritz) to the snout of the nearby glacier, Vadret da Morteratsch, in the Engadin. Another example is the Rhonegletscher, the source of the River Rhône in the Valais, which can be reached by road. At both places, visitors may observe the rate at which plants have recolonized terrain that became ice-free only years or decades ago. How did the climate change in order to cause such a dramatic glacier recession? Temperature records from Basel in Switzerland, which began in 1755, show that cold summers were much more

Receding glacier, Steingletscher, Sustenpass, Switzerland. The photos were taken in 1987, 1996 and 1999 from exactly the same location. Note how the glacier not only recedes but also how its surface becomes progressively flatter. A smooth gently sloping front and relatively few crevasses are typical of such temperate glaciers in a state of recesssion.

common during the Little Ice Age of the mid-eighteenth and mid-nineteenth centuries. Summer temperatures affect glacier melting, and are therefore commonly much more significant for a glacier's mass balance than are winter temperatures. At the same time precipitation did not decrease. Thus temperature seems to have been the major controlling climatic factor. Even so, the size of these temperature changes was surprisingly small (one or two degrees Celsius of summer cooling) considering that the recession of a typical glacier tongue in the Alps has subsequently been typically as much as a kilometre. These historical records reveal that glaciers in temperate regions are extremely sensitive to temperature fluctuations, and can therefore be used to reconstruct climatic trends in areas with few or non-existent climatic data.

Tidewater glaciers fluctuate rapidly, and it is sometimes possible to find both advancing and receding glaciers side-by-side in areas like Alaska. Here Cascade Glacier in the Barry Arm of College Fiord, southeast Alaska shows the dark stripe of a prominent medial moraine as it terminates in a cliff grounded on the sea bottom.

Glaciers in the Alps, as in many parts of
the world, have receded dramatically
from their most historically recent
maxima, which occurred in a period
known as the Little Ice Age (around
1750–1850). This painting by Birman in
1826 illustrates the Mer de Glace near
Chamonix, France soon after the glacier
reached its maximum extent. It is
interesting to compare this painting with
the photograph on page ii, as the surface
of the glacier has dramatically lowered
during the intervening period. (From the
Gugelmann Collection, Prints and
Drawings Department, Swiss National
Library, Bern.)

The 1950s brought fears that many Alpine glaciers might disappear altogether, following the long period of slowly rising temperatures since the Little Ice Age. However, a slight cooling then set in, which lasted into the 1970s, and for a few years more of the regularly documented Swiss glaciers were advancing than were receding. In fact, some Alpine glaciers responded with positive gusto to the new cooling phase; for example, the Allalingletscher advanced 174 metres in 1970–71 and the Oberer Grindelwaldgletscher 100 metres in 1971–72. However, the response time of some of the larger Alpine glaciers has been too slow to take advantage of this cool phase, which in any case only proved to be short-lived. For example, the reaction time of the Grosser Aletschgletscher is so long that it failed to stop receding before the succeeding warmer phase of the 1980s and 1990s. The temperatures have risen so much during the past two decades that the majority of Alpine glaciers are again undergoing rapid recession.

The most easily visited part of a glacier for walkers is usually the snout. Here, an enthusiastic observer may, over several years, enjoy undertaking a simple but rewarding photographic project. Photographs taken from the same spot, especially of a side view, will

Figure 4.1 The terminal positions of many glaciers in Switzerland have been recorded annually for over a century. Three examples are shown here, the vertical axis showing in metres the cumulative recession since 1895. Both the Grosser Aletschgletscher (red line) and the Vadret da Morteratsch (purple line) show continuous recession, amounting to around 2000 metres. In contrast, Steingletscher (green line) reveals two main periods of readvance, in 1910–25 and 1970–90, but the net result is still recession, in this case of around 600 metres.

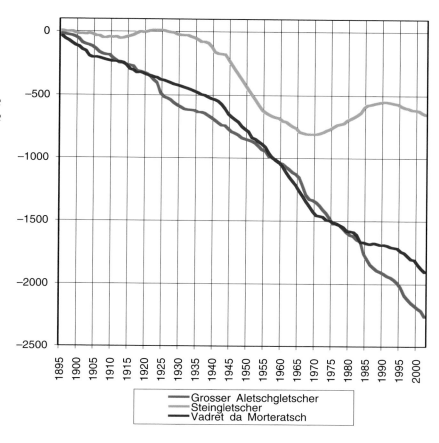

Overleaf Tidewater glacier cliff reflected in a pool on melting sea ice, Nordenskiöldbreen, Billefjorden, western Spitsbergen.

clearly document the recession or advance of a glacier over the course of a year. However, even by visiting a glacier only once it is usually possible to judge whether it is advancing or receding.

Receding glaciers have a gently graded, flat snout, often allowing fairly easy access onto the ice. Large areas of stagnating debris-covered ice give rise to unstable and hummocky topography. A meltwater stream may well emerge from a wide-open ice cave or glacier portal. Furthermore, there will be no plants on the ground adjacent to the ice since it will have only recently been uncovered.

An advancing glacier snout, in contrast, usually has a steep convex front which may be difficult to scale except by experienced ice climbers. The meltwater stream usually emerges from the ice without creating a cave, since the advancing ice squeezes shut any opening. During the short advance phase of the late 1970s, several

Alpine glaciers reached terrain that had been ice-free for several decades, so that trees had had time to establish themselves. Seeing trees being pushed over by a glacier is a dramatic illustration of glacial advance, but is a rare sight in the Alps today.

Even though many glaciers show year-by-year recession cycles, small winter advances may take place when there is little melting. This behaviour is manifested by a set of small annual **push moraines**. A series of small push moraines up to a metre high can be observed near the snout of Glacier de Tsanfleron in Switzerland, for example.

The variable nature of glacier advance and recession since the Little Ice Age may seem somewhat random. In fact, fluctuations are influenced mainly by glacier size and shape. Four main types of behaviour have been identified. Firstly, large valley glaciers such as Gornergletscher have undergone rapid, continuous recession. Secondly, small mountain glaciers, such as Glacier de Saliena, have shown pronounced recession interrupted by two twentieth century advances. Thirdly, large mountain glaciers, such as Vadret da Tschierva, are similar to the second group but show more significant recession. Lastly small cirque glaciers such as Glacier de Plan Nevé have shown slow recession through the twentieth century. A similar pattern of behaviour has been documented in the Cascades of Washington State.

Tidewater glaciers

Glaciers terminating in the sea behave differently from those on land. Instead of slowing down towards the snout, the ice accelerates as it enters the sea, resulting in a glacier surface that is almost impossible to traverse because of extensive crevassing. The same is also true for glaciers ending in deep lakes. Tidewater glaciers are typical of the fjords of Alaska, Chile, Svalbard, the Canadian Arctic, Greenland and the Antarctic Peninsula. For the temperate glaciers of Alaska and Chile, the terminus normally is a vertical cliff, grounded on the sea bottom. The cliff may extend at least 100 metres below sea level and rise 50 or more metres above the water. The extensive crevassing and forward motion of the ice produces a very unstable face, with the result that large pinnacles and enormous blocks of ice frequently topple and crash into the water, creating magnificent displays as fountains of water shoot high into the air.

Sometimes calving occurs from the ice cliff below water level, giving rise to the unnerving sight of blocks of ice being projected up out of the water. It is accompanied by unpredictable waves, which can come as a shock to anyone camping close to the shore, or sailing too close to the glacier. The blocks disintegrate on hitting the water, and the resulting icebergs are rarely more than a few tens of metres across. Many smaller pieces are produced, often termed **bergy bits**. All this activity is a feature of temperate and many polythermal tidewater glaciers, irrespective of whether they are advancing or receding. A temperate glacier terminus is usually too weak to float on water, so floating valley glacier tongues are restricted to polythermal glaciers in which much of the ice is below the melting point. An

Fast-flowing glaciers in Greenland commonly discharge into fjords too deep for the ice to remain grounded on the bed, so they begin to float. Large tabular icebergs are produced by such glaciers, as here at the head of Nordvestfjord in East Greenland.

additional requirement is water several hundred metres deep, as can be found in the fjords of Greenland. Here, the lower reaches of the glacier lift off the bed and begin to float at the so-called **grounding line** or **zone**. Large **tabular icebergs** up to several hundred metres in length calve from these floating glacier termini.

Nowadays, most tidewater valley glaciers terminate well inside the fjords that were created when ice reached the open sea. They have receded rapidly since the Little Ice Age of the eighteenth and nineteenth centuries. Only a few icebergs reach the open sea, and many become stranded on banks, especially in areas of high tidal range. Here, they leave grooves and push ridges as they are dragged across the sediment.

Icebergs have a particularly interesting property. As glacier ice melts in water, the air bubbles trapped under pressure for hundreds of years are released, making a crackling, popping noise. In North America this noise is referred to as ice sizzle or bergy selzer. Some scientists on fieldwork occasionally like to add a piece of an iceberg to a 'wee dram' of Scotch! The release of the pressurised air creates a fine pleasurable spray as one drinks the whisky.

Fluctuating tidewater glaciers in Alaska

The most intensive studies of tidewater glaciers have been along the south coast of Alaska. Here, most glaciers have receded huge distances in 200 years, yet some occasionally readvance. Reduced ice supply from the accumulation area has led to the recession of several very long valley glaciers that terminate in fjords. Fjords are commonly deepest in their inner reaches because erosion by ice has been at its greatest there, so that as the glaciers recede into deeper water and contact with the bed is weakened, their recession becomes more pronounced.

Glacier Bay
Glacier Bay, a fjord system up to 550 metres deep in the Panhandle of Alaska, displays a remarkable record of glacial recession over the

last two centuries, combined with rapid colonization by a succession of different plant species. In 1794, near the end of the Little Ice Age, the British Navigator George Vancouver observed at the mouth of what is now Glacier Bay 'an immense body of compact perpendicular ice, extending from shore to shore, and connected with a range of lofty mountains on each side'. The first scientific exploration of Glacier Bay, in October 1879, was by John Muir, the famous Scottish naturalist and mountaineer, who had emigrated to the USA and led the movement to establish the first national parks in his adopted country. This is how he described his first view of this majestic fjord.

I therefore set out on an excursion, and spent the day alone on the mountain-slopes above the camp, and northward, to see what I might learn. Pushing on through rain and mud and sludgy snow, crossing many brown, boulder-choked torrents, wading, jumping, and wallowing in snow up to my shoulders was mountaineering of the most trying kind. After crouching cramped and benumbed in the canoe, poulticed in wet or damp clothing night and day, my limbs had been asleep. This day they were awakened and in the hour of trial proved they had not lost the cunning learned on many a mountain peak of the High Sierra. I reached a height of fifteen hundred feet, on a ridge that bounds the second of the great glaciers. All the landscape was smothered in clouds and I began to fear that as far as wide views were concerned I had climbed in vain. But at length the clouds lifted a little, and beneath their grey fringes I saw the berg-filled expanse of the bay, and the feet of the mountains that stand about it, and the imposing fronts of five huge glaciers, the nearest being immediately beneath me. This was my first general view of Glacier Bay, a solitude of ice and snow and newborn rocks, dim, dreary, mysterious. I held the ground I had so dearly won for an hour or two, sheltering myself from the blast as best I could, while with benumbed fingers I sketched what I could see of the landscape, and wrote a few lines in my notebook. Then, breasting the snow again, crossing the shifting avalanche slopes and torrents, I reached camp about dark, wet and weary and glad.

From Muir, J., *Travels in Alaska*
reprinted in *The Eight Wilderness Discovery Books*, 1915, London. Diadem
Books; Seattle: The Mountaineers, 1992.

By the time of Muir's visit the ice had already receded over 60 kilometres up the bay from its Little Ice Age maximum, and the bay was full of icebergs. By the end of the twentieth century, the recession was approaching 100 kilometres, and relatively few icebergs were being produced. Furthermore, many of the former tidewater glaciers have now receded onto land, including that named after Muir himself.

Most glaciers in Glacier Bay continue to recede at a rapid rate, but some fed by the higher mountains of the Fairweather Range and the St. Elias Mountains were advancing until late into the twentieth century. One glacier caught in this ebb and flow, the Grand Pacific, is unsure about its national identity! Beginning in the USA (Alaska), it flows through Canada (the Yukon) and terminates in Glacier Bay close to the international boundary, which from time to time it crosses back into the USA. It currently terminates in Alaska.

The phenomenal recession of the glaciers in this area has been accompanied by some of the most active geological processes on Earth. Several factors are involved, including collision between the Pacific Ocean plate and the North American continental plate, upward readjustment of the land following removal of the ice burden, and even greater rapid crustal uplift in this earthquake-prone region. At Bartlett Cove, near the mouth of the bay, the land is rising at a rate of four centimetres a year, having risen several metres since the 1790s. A combination of unstable land, loose glacial debris and high precipitation also leads to rapid erosion and deposition, and some of the newly vacated fjords are filling with sediment at a rate of several metres a year.

Colonization by plants is equally rapid, and each succession makes a contribution to the building of the Sitka spruce forest now developed on the Little Ice Age moraines. As the ice recedes from the land, the first plants to appear are lichens and mosses. Mat-forming *Dryas* and other low-growing plants follow. These give way in turn to alder, willow and cottonwood, and finally to Sitka spruce. The climax of the succession has yet to be reached in Glacier Bay, but a few western hemlocks are beginning to make their appearance in the region of Bartlett Cove. The inner branches of the bay recently vacated by ice provide evidence that coniferous forest grew all over the region prior to the Little Ice Age. Trees pushed over thousands of years ago by advancing ice have been preserved in glacial sediments and are now emerging in areas prone to erosion.

Prince William Sound

At the head of Prince William Sound in south-central Alaska is the spectacular calving tidewater Columbia Glacier. It is claimed that this is now the fastest glacier in the world. This glacier has undergone a substantial change in behaviour over the last two decades. Up to 1982, the glacier terminus was relatively stable as it was anchored on a rock barrier or **sill** but, as it thinned, the ice in the deep basin behind became buoyant. This thinning proved to be the trigger for rapid recession and rapid acceleration of velocity from 5 to 15 kilometres a year. By 2001 the glacier had receded 13 kilometres, reducing its length to 54 kilometres. Recession into progressively deeper water (now 200 metres) has been accompanied by increasing iceberg production, from 3 to 18 million cubic metres a day.

The future prospects of Columbia Glacier are not good. As the glacier recedes, the water in the lengthening marine basin will deepen to around 700 metres, rendering the tongue increasingly unstable. It is only likely to stop receding when the terminus emerges from the sea. This type of behaviour, albeit exceptionally dramatic, is typical of tidewater glaciers filling deep marine basins: once they become sufficiently thin, they start to float and break up rapidly. This is probably what John Muir was observing in Glacier Bay in the late nineteenth century.

The behaviour of Columbia Glacier is important in economic terms, as the increased iceberg production is affecting shipping lanes to the port of Valdez. Indeed, it has been reported that when the tanker *Exxon Valdez* ran aground on Bligh Reef in Prince William Sound on 24 March 1989, it was trying to avoid an iceberg. The accident led to the spillage of 11 million gallons (42 000 cubic metres) of Alaskan crude oil. Within days approximately 1100 kilometres of coastline was polluted, ultimately killing thousands of sea otters and hundreds of thousands of seabirds. By the end of the century just two species of wildlife, bald eagles and river otters, were said to have recovered from the accident. The impact on the local fishing economy has also been very severe.

Yakutat Bay

In contrast to the recent rapid recession of tidewater glaciers in other parts of Alaska, the huge glacier complex dominated by the Hubbard Glacier in Disenchantment Bay in the inner reaches of Yakutat Bay has been slowly advancing since 1894 when it was first surveyed. Today, Hubbard is the longest valley glacier in North America, extending approximately 123 kilometres from its source in Canada. Its recent behaviour has been dramatic, blocking the entrance to Russell Fiord twice in the last two decades, creating large temporary lakes. There was considerable concern amongst the local communities as it became apparent that overflow might occur at the opposite end of the lake to the ice dam, flooding settlements including Yakutat.

The first such event took place between May and October 1986, when possibly as a result of a surge of Hubbard Glacier's tributary, Valerie Glacier, the mouth of Russell Fiord, was dammed. The water level in the resulting Russell Lake rose by 25 metres, flushing out nesting birds, and trapping fish and seals, but not reaching a level that threatened flooding of the communities around Yakutat. The dam eventually failed and the lake emptied within just 24 hours. Glaciologists calculated that the outburst from this glacier-dammed lake was the biggest in North America since the last ice age ended, some 10 000 years earlier.

The second event took place in summer 2002 when Hubbard Glacier advanced once again across the mouth of Russell Fiord. By late June the lake was already forming when the US Geological Survey installed a water gauge. On this occasion the advancing glacier pushed up a terminal moraine of loose marine sediments to block the mouth of Russell Fiord. After continuing to build up to a level 20 metres above sea level, on 14 August the dam once again failed. A raging torrent of water and icebergs now connected the lake to the sea, and the lake emptied within 36 hours. The Survey glaciologists estimated that, at its peak, discharge was 30 times the maximum flow of the Mississippi River at Baton Rouge in recorded history, although much less than that of the 1986 outburst. Once again the threat to Yakutat was averted. Throughout the blockage period, there was a delicate balance between the advance of the glacier, the build-up of the moraine, and erosion of the dam by water.

Scientists predict that the dam will re-form as the glacier continues its inexorable advance. Its terminus is moving forward at 30 metres per day, and will eventually overcome the powerful tidal currents that at present keep the channel open.

Small-scale advances of otherwise receding glaciers occur in some instances. In this view of the snout of Mackay Glacier, western Ross Sea, Antarctica, an unusual example shows sea ice pushed up as curving overlapping slivers as seen towards the end of the winter season. These delicate features will soon disintegrate as summer arrives.

Kilimanjaro (5895 metres) is a volcano and Africa's highest peak. It carries remnants of a glacier that is in danger of disappearing totally within the next two decades. Vertical ice cliffs along the glacier margins are a result of intensive solar radiation (Photograph courtesy of Walter Hauenstein).

Opposite. Artesonraju, dominated by the peak of Nevado Pirámide, is a glacier that terminates at the edge of a cliff, below which is now a lake. Peruvian glaciologsts are undertaking mass-balance studies on this glacier, which is one of the few accessible clean glaciers in the Cordillera Blanca, and thus provides a clear signal of the response of these glaciers to climatic change in the tropics.

Recession of tropical glaciers

Of all the world's glaciers, those in the tropics have seen the greatest percentage loss of area over the last century, as the following statistics from a report in 2002 indicate:

Kilimanjaro Ice Cap, central Africa:	82% loss since 1912
	33% loss since 1990
Ruwenzori glaciers, central Africa:	70% loss since 1906
Mt. Kenya glaciers, central Africa:	40% loss since 1963
Quelccaya Ice Cap, Peru:	20% loss since 1963
Glaciers in Venezuela:	6 glaciers in 1972 reduced to 2

All these glaciers are located on mountains 5000 to 7000 metres high, and many of them will disappear within the next few decades as a result of climatic warming. The loss of tropical glaciers will be unfortunate for the local communities, as some rely heavily on

glacier meltwater for cultivation, drinking and hydro-electric power generation. La Paz in Bolivia (population 1.7 million) depends on meltwater from glaciers in the Cordillera Real. Planning for replacement sources is urgently needed in these areas. On a local scale, the village communities in the Cordillera Blanca in Peru have constructed an intricate network of irrigation channels, and are thus able to grow crops and support animals in an area that lacks precipitation for half the year.

Human-induced glacier recession

Humans have modified glaciers in a variety of ways. Most obvious is where glaciers have been induced to recede by creating lakes, and

Floating tidewater glaciers occur in areas where the mean annual air temperature is well below freezing. Instead of small, irregularly shaped icebergs, such glaciers produce tabular icebergs. Some tabular icebergs are seen here in various stages of formation and decay at the terminus of a glacier in Whisky Bay, James Ross Island, Antarctic Peninsula. In the background are small stagnant glaciers wasting away *in situ*, and illustrating the rapid climatic warming that has been taking place since the 1950s in this part of Antarctica.

A rapid advance of a glacier front can be triggered by a surge of one of its tributaries, as when Valerie Glacier surged into Hubbard Glacier, southern Alaska in July 1986. A lobe of Hubbard Glacier, which generally forms a regular calving cliff line, extended across a relatively shallow stretch of sea bottom and blocked the mouth of Russell Fiord. This arm of the sea thus became a huge lake which over the course of four months rose by 30 metres, flooding the forests adjacent to it, trapping seals and displacing nesting birds, until the ice dam constraining the lake broke catastrophically. The event caused much concern because it was feared that the village of Yakutat and its airport could be flooded. A similar blockage and dam failure also occurred in 2002.

Pastaruri is a small rapidly disappearing icefield near Huaraz in the Peruvian Andes. Its summit is only just above 5000 metres and at current rates of recession is likely to disappear within 20 years. At this altitude glaciers are now unsustainable in the tropics.

allowing the glaciers to calve into them. Dams for hydroelectric power generation were constructed in front of Ghiacciao del Sabbione and Griesgletscher, two Alpine glaciers that have adjacent accumulation areas on the Swiss-Italian border. Filling of the resulting lakes led to a rapid recession of the ice. This had been taken into account when planning the storage capacity of the reservoirs. A similar, but rather bigger project was evaluated in the Bernese Oberland of Switzerland, where a large dam was planned that would have flooded Unteraargletscher and shortened it by three or four kilometres. However, concern for the outstanding Alpine scenery eventually led to the abandonment of this scheme.

Human civilization is also contributing to glacier recession indirectly – through generating greenhouse gases such as carbon dioxide and methane that lead to warming of the atmosphere. Although it is still difficult to differentiate between human-induced climatic changes and natural warming since the Little Ice Age, there is an increasing body of evidence to suggest that glacier recession is accelerating as the burning of fossil fuels intensifies. These are issues of global importance, and are addressed more fully in Chapter 16.

Using the latest projections of climatic warming trends into the next few decades, it appears reasonable to expect an increase in the equilibrium line altitude in the Alps of about 300 metres by the year 2035. In Switzerland this would mean that about half of all existing glaciers would disappear by then. Naturally, many small mountain glaciers would cease to exist first. Larger glaciers would remain but recede rapidly up alpine valleys. It seems clear that we shall have to face drastic changes in the appearance of glacierized mountain regions in the coming decades.

5 Ice on the move

When scientists first studied glaciers in the early nineteenth century, the complexities of glacier flow were unknown. Combining mountaineering with science, people such as James Forbes and John Tyndall set about making measurements of the flow of valley glaciers such as the Mer de Glace in France and the Unteraargletscher in Switzerland. Some of the fundamental aspects of glacier flow were determined at this time. Today we recognize that there are a wide range of features that indicate glacier flow. The formation of crevasses and other structures, the displacement of rocks on the surface and the occasional cracking and creaking sounds within the ice are all symptoms of this. The eroded rocks and deposits that are left behind after the glacier has receded also indicate how glaciers move.

Rates of movement of flowing glaciers are extremely variable. Some small glaciers and ice caps may flow only a few metres a year. On the other hand, the fastest part of an average-sized valley glacier flows typically between 50 and 400 metres a year, even several kilometres if they end in the sea. Similarly, large ice streams that drain the Antarctic and Greenland ice sheets flow steadily at rates of several kilometres a year.

A small percentage of the world's glaciers flow in a rather unpredictable manner; they remain relatively inactive for many years, but may accelerate suddenly, increasing many hundredfold in speed. For a period of a few months or years they advance over distances measured in kilometres, in what are known as **surges**, phenomena that have given rise to the colloquial term 'galloping glaciers'.

The midnight Sun illuminates two glaciers on Axel Heiberg Island in the Canadian Arctic, with White Glacier in the foreground and Thompson Glacier in the background. The low light emphasizes surface features such as crevasses that indicate flow towards the right of the photograph.

How do glaciers flow?

Ice flows by three main modes: by internal deformation (or creep), by sliding over a hard rocky bed and by movement over a soft deformable bed of sediment.

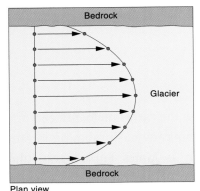

Plan view

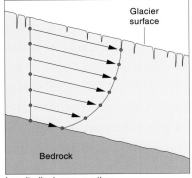

Longitudinal cross-section

Figure 5.1 Two of the components of glacier flow in a valley glacier, both in plan view and longitudinal cross-section. The markers are usually stakes on the surface and a borehole from surface to bed. The curved lines represent displacements in the direction of the arrows, i.e. internal deformation, and shows faster movement in the middle and upper reaches of the glacier. The displacement against the bedrock is basal sliding.

Internal deformation

As snow turns to firn and then ice, its constituent crystals alter under the weight of material as it is buried, and is subject to the influence of gravity. The resulting stresses cause the ice at depth to deform in a rather plastic manner, much as soft putty does when kneaded, only considerably more slowly. Deformation is greatest near the bed and sides of a glacier, so a typical flow pattern shows an initial rapid increase in velocity away from the margins, then a declining rate of increase towards the middle, as the diagram illustrates. Such a velocity profile is parabolic in shape. Similarly, there is a rapid increase in velocity in the first several metres above the bed of a glacier, and then the rate of flow increases only slightly above this basal zone.

Glacier flow has a profound effect on the nature of the ice. The upper layers – down to depths of 30 metres or so in temperate glaciers, and more in cold glaciers – are brittle under tension. When they move they crack, creating one of the most hazardous features of glaciers – the open fractures called **crevasses**. Less dramatic, but equally characteristic, are a wide range of other layered structures, reflecting deeper plastic deformation.

Basal sliding

The second component of glacier flow is basal sliding, where the glacier slips over a hard rocky bed. Large quantities of meltwater produced in summer reduce the friction between the glacier and its bed, resulting in faster flow. In a temperate glacier, basal sliding is the principal component of flow, and may account for as much as 90 per cent of its overall movement. Where basal sliding occurs over uneven bedrock it often generates caves where glaciologists have been able to observe at first hand the range of processes of erosion and deposition – albeit in uncomfortable and not always safe circumstances!

Since sliding velocities are related to the amount of meltwater available, a temperate glacier moves faster in summer than in winter, and faster in daytime than at night. Exceptional speeds may also be induced by heavy rain. In cold glaciers basal sliding can only occur where the ice is thick enough for the base to be warmed to melting point by geothermal heat. Moreover, it is quite common for the snouts of otherwise sliding glaciers to be frozen to the bed because the ice is thinner there, and thus affected more by the low mean annual air temperature.

Basal sliding produces several phenomena, including fine sediment suspended in meltwater (**glacier milk**), marks on the bedrock such as scratches called **striations**, and a distinctive type of sediment, **till** – all of which are described in Chapters 6 and 10.

Movement over soft, deformable beds

A layer of unconsolidated sediment known as till frequently underlies moving ice. This is a mixture of particles of all sizes from clay to boulders. 'Till' is an old Scottish term for stiff stony ground, but is now applied internationally to deposits released directly from glacier ice. When saturated with water, this sediment deforms more easily than the basal ice, and glacier movement is assisted by shearing within the soft, deformable sediments, rather than by sliding. A wide range of tests need to be applied to the sediment in order to establish whether the ice was moving over a deformable bed, including stone-orientation measurements, stone shape, grain-size distribution, internal layering and shear strength.

Structures in glacier ice

A glacier may be considered as a small-scale model for the processes that occur deep in the Earth's crust, such as those displayed in the Rocky Mountains or the Alps, as a result of the collision of continental plates. If you walk over bare glacier ice or peer down into crevasses you can see a wide range of layered structures. Like rocks in mountain ranges, the layers may be continuous or discontinuous, folded, or form complex patterns. All result from glacier flow, and mostly they reflect the plastic behaviour of ice deep in the glacier. On the smooth surface of a glacier after rain they appear as beautiful layers of contrasting colour and texture, from blue to white, and comprising coarse-grained to fine-grained ice crystals. After a period of weathering, especially under the influence of solar radiation, the surface may become furrowed, reflecting the faster rates of melting of the darker ice layers.

Accumulation layering

If we examine a glacier from its upper reaches to its snout, we can determine how the different layered structures evolve from the initial snowfalls in the accumulation area. After the year-by-year accumulation of snow and its subsequent change, first into firn, then into ice, a layered structure called **sedimentary stratification** develops. Each annual accumulation is represented by a thick layer of light-blue, coarse-grained bubbly ice up to a few metres thick, separated by thin layers of dark blue, coarse-grained clear ice. The former is the result of direct conversion of snow to ice by pressure under the

Svinafellsjökull is an outlet glacier from Oræfajökull ice cap in southern Iceland. Pouring over a high rock step, the ice generates curved structures called ogives that indicate how the ice flows faster in the middle of the glacier than at the sides. Winter and summer flow appear to generate pairs of light and dark ice respectively. It is therefore easy to estimate the glacier's ice velocity below the icefall.

increasing weight of snow above, whereas the latter reflect local horizontal zones or layers, which had become saturated with water during the melt season and subsequently refroze.

As the glacier moves downhill the layers deform gently, because ice is approximately plastic and moves faster in the middle of the glacier than at the margins. A period of excessive ablation may remove and truncate many of the layers, so that when new layers accumulate a marked discontinuity, called an **unconformity**, is apparent.

Folds and foliation

As flow continues, stratification and other ice layers become more tightly folded and sheared, giving rise to a new layered structure called **foliation**. Both **folds** and foliation generally develop at depth in the plastic flow zone.

Fast-flowing tidewater glaciers are especially prone to heavy crevassing. Kronebreen, NW Spitsbergen, in this view flows at 700 metres per year, and its surface is a maze of crevasses, making cross-glacier travel impossible. At the bottom of the picture, its tributary Kongsvegen is moving at less than 10 metres per year, as reflected in the absence of crevasses.

A fine set of medial moraines on Edward Bailey Gletscher, Milne Land, East Greenland. Note how convergence of two streams of ice with debris along the sides (as lateral moraines) combines to produce medial moraines. Debris commonly extends down to the base of the glacier in these moraines, and the associated ice is very strongly sheared and folded. The glacier is flowing towards the bottom of the picture.

Foliation superficially resembles stratification, but the layers are thin and discontinuous. The structure is most strongly developed where shearing is greatest, such as close to the glacier margin, or where two streams of ice combine. In such cases, the coarse ice crystals may break down into fine-grained ice of whitish, granular appearance. Normally, foliation of this type is parallel to the direction of flow. In thick mist we have occasionally found this a useful way to keep our bearings on a flat glacier, the structure showing us the way up or down. As ice moves down-slope, new foliations overprint older ones, so the pattern exposed at the glacier surface may be exceedingly complex.

A different style of folding may be evident if the glaciers have margins with cliffs, as is common in the Polar Regions. These folds commonly show up well, as layers of debris enhance them. Such

Most glaciers develop a prominent layered structure called foliation. This structure is the product of strong deformation such as shear or compression. Being made up of different types of ice (bubble-rich and bubble-poor), it is subject to differential weathering, producing a characteristic furrowed surface. This fine example occurs on a glacier named Comfortlessbreen at Engelskbukta, western Spitsbergen.

folds are recumbent, that is they lie on their side from the dragging effect of ice moving over an uneven bed.

The basal ice layer

The base of a glacier consists of a zone of layered ice and debris that typically ranges in thickness from a couple of metres or less in temperate glaciers, to tens of metres in polythermal glaciers. Many glaciologists refer to this **basal ice layering** as 'stratified ice'. It results from the freezing on of subglacial water and debris from the glacier sole, which is then subject to strong shear. The ice is quite different from bubbly glacier ice derived from snow: it is clear and dark in appearance. Debris occurs in several forms: as discrete layers, disseminated through the ice, or as muddy clots. People may best observe the basal ice layer in caves beneath the glacier, although such sites can be dangerous, as stones or blocks of ice may fall from the ceiling.

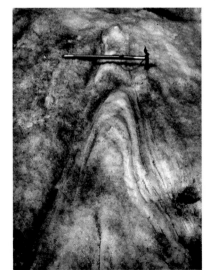

At depth within the glacier, ice deforms rather like a plastic material such as plasticene. Existing layers can become folded as at the surface of Griesgletscher in the Swiss Alps.

Veins related to fracturing

We may observe other sorts of layers at the glacier surface that result from the fracturing and stretching of ice, but without necessarily forming open crevasses. Near-vertical faults form by displacements

Folding of ice is seen well in ice cliffs, especially if debris from the base is entrained as well. This Z-shaped fold is in the vertical cliff of Thompson Glacier on Axel Heiberg Island in the Canadian Arctic.

along fractures both vertically and laterally. Low-angle faults called **thrusts** may generate at the glacier bed and extend upwards and forwards, such as where the ice is slowing down. Such features are also accompanied by folding in many cases, and are especially common in polythermal glaciers where the conditions at the bed change in a down-glacier direction from basal sliding to frozen. Thrusts may

During a surge (a period of exceptionally fast flow) of Variegated Glacier in Alaska, the ice is not only folded, but also faulted. The diagonal lines from bottom right to top left are low-angle faults called thrusts, where the ice up-glacier overrides the ice below. Folds and thrusts are well-exposed in this image, taken soon after the 1982–83 surge when the surface was still very rough because of the extensive crevassing that occurred at the time.

carry a large amount of basal debris into the body of the glacier, and some of it may be raised to the surface.

Many crevasse sets are associated with veins of clear blue ice. These are either the traces of old, water-filled crevasses, or are formed as a result of recrystallization of ice in narrow zones under tension. These features, known as **crevasse traces**, develop with continued flow into a different sort of foliation from the flow-parallel variety described above. Crevasse traces are most commonly found as transverse sets that develop into a curved structure known as arcuate foliation. Near the margins of the glacier the two types of foliation may merge into one another because the faster flow in the middle of the glacier produces crevasse traces with longitudinal orientation near the margins.

Glaciers that flow over icefalls commonly produce structures called ogives – surface waves at the base, followed by curving light and dark bands reflecting the way the ice flows. Many scientists believe that they are produced annually. These ogives are at Bas Glacier d'Arolla in Valais, Switzerland, but unusually are not being produced annually at the present time.

Ogives

Among the most striking of all glacier structures are **ogives**. These are bands that curve across the glacier surface and originate within and below the chaos of collapsing blocks in an icefall. They are made up of sets of light and dark bands or waves, each usually several metres wide. Ogives only develop within icefalls but, for some reason, not all icefalls generate ogives.

In detail, the individual light and dark bands each comprise many layers similar to arcuate foliation, but the proportions of different ice types vary. Glaciologists have shown that at least some ogives are annual features, a pair of light and dark bands or a wave representing a year's movement through the icefall. The ice, being thinner in summer, produces a trough, while the thicker ice of winter produces the crest of a wave. Thus by measuring the distance across a pair of light and dark layers we can roughly determine the ice velocity.

Band ogives, also known as Forbes' bands after the nineteenth century Scottish glaciologist who first described them, reflect the passage of dirty ice in summer and snow-covered ice in winter through the icefall, hence the light and dark bands. According to some glaciologists, wave ogives reflect the passage of thinner ice (because of ablation) through the icefall in summer, but strong deformation where the ice rapidly slows down at the foot of the icefall may also play a role. Recent research has shown that, since basal debris is sometimes associated with ogives, folding and thrusting of the basal debris layer may also have a role in ogive formation. Both wave and band ogives are surprisingly persistent, and commonly can be traced all the way to the snout. Those on the Mer de Glace in the French Alps are perhaps the best known, as many visitors who take the funicular railway from Chamonix to Montenvert can view them.

Distance between the bands is not the only guide to glacier speed: by counting the number of ogives we can determine how long it has taken ice to cover the distance from the icefall to the snout, as in the Mer de Glace, which has about 50, thus indicating about 50 years.

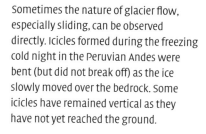

Sometimes the nature of glacier flow, especially sliding, can be observed directly. Icicles formed during the freezing cold night in the Peruvian Andes were bent (but did not break off) as the ice slowly moved over the bedrock. Some icicles have remained vertical as they have not yet reached the ground.

Glaciologist measuring deformation in the basal layers of the Glacier de Tsanfleuron, Switerland using a sophisticated electronic instrument. In the foreground is wet, striated rock (limestone) over which the ice slides. Subglacial caves like this, near the receding ice margin, rapidly disappear, but new ones continually form where there are steps in the bed.

Moreover, as ogives move down-glacier they often become more closely spaced, indicating that the ice is slowing down.

Crevasses

Crevasses are V-shaped clefts that may be many times deeper than they are wide, and are among the most perilous hazards faced by the mountaineer. Many crevasses are covered by bridges of snow that hide them from unwary glacier travellers. Many climbers and walkers have fallen to their deaths when the snow bridges collapsed as they unwittingly tried to walk across them. Under good lighting conditions you may be able to recognize a bridged crevasse from the subtle shades of reflected light on the snow surface. However, in the flat light under cloudy skies there may be no visible sign that the area is crevassed. Furthermore, drifting snow may hide crevasses so well that even in clear weather they cannot be recognized even by an experienced mountaineer. Thus, you have constantly to probe the snow with an ice axe if you are to remain safe whilst walking across a glacier surface.

Newly formed crevasses are generally clean-cut and straight-sided. As crevasses get older and the ice melts, their walls become less steep and more rounded, and in this state the walker wearing

An aerial view of the heavily crevassed accumulation basin of Fox Glacier, South Island, New Zealand is typical of glaciers with heavy snow precipitation as this induces fast flow. Apart from those crevasses that are visible, many others are bridged by snow, so walking across such terrain would be very hazardous.

crampons may walk into them without difficulty, although some may contain ponds. Eventually, melting may remove most traces of these hazardous obstacles to travel.

Rescue from freshly formed crevasses in the accumulation area may be difficult, since they have overhanging sides and unstable bridges. Considerable expertise in handling ropes is essential for anybody crossing fields of snow-covered crevasses. However, it is unlikely that anyone falling into a crevasse would actually plummet to the bottom, because they usually contain old collapsed snow bridges that will break a fall.

As the ice moves, crevasses close and there have been instances in the European Alps of unrecovered bodies becoming embalmed in the ice, only to be released several decades later. One well-known

Crevasses are associated with 'extending flow', that is a zone of accelerating flow, as here in Glacier de Saleina, Valais, Switzerland. These transverse crevasses form as thin cracks that widen progressively. Then, as the ice approaches the lip of an icefall, the blocks between the crevasses begin to collapse.

case took place in 1820 when three guides in a climbing party were swept by an avalanche into a deep crevasse on the Glacier des Bossons on the slopes of Mont Blanc near Chamonix in the French Alps. It was only after 43 years that the bodies were released from the ice near the glacier snout, having travelled just over three kilometres in that time.

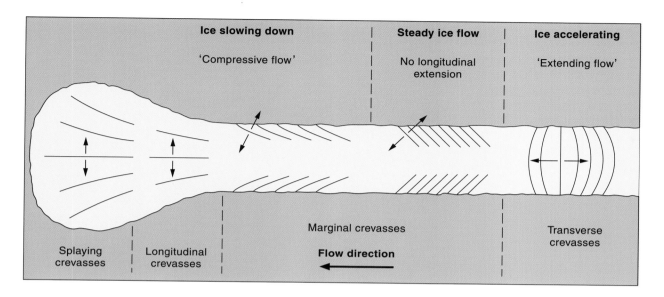

Figure 5.2 Plan view of the principal types of crevasse in the tongue of a valley glacier, together with the types of flow involved. The arrows, which are at right angles to the crevasses, indicate the directions in which the ice is pulled apart.

The mystique of crevasses and their frequent appearance of being bottomless has led to many exaggerated estimates concerning their depths. Claims that crevasses are hundreds of metres deep are not unusual, yet in the temperate glaciers typical of alpine regions they are rarely open to depths of more than a few tens of metres. On the other hand, the cold ice sheet of Antarctica has huge crevasses that could easily swallow an object as big as a London bus, let alone over-snow vehicles such as snow scooters. However, few reliable measurements of their depth exist.

The principal reason why crevasses are not 'bottomless' is that only the upper part of the glacier is brittle. Below a critical depth (around 30 metres in temperate glaciers) the weight of overlying snow and ice makes the ice more plastic. Thus, any fracture that propagates down from the surface cannot open up below that depth, because stretching becomes counter-balanced by plastic deformation. The exceptions to this rule occur if there is already a weakness in the ice, say from an older fracture formed further up the glacier, or if the crevasse is filled with water.

Crevasses form where the ice is under tension, such as where a glacier flows over a step in the bedrock, round a bend, or where the valley that constrains it narrows or widens. They normally form in

Glacier des Bossons in the French Alps, descending from Mont Blanc, is one of the steepest glaciers in the region. Fed by heavy snow on this lofty peak, the glacier flows rapidly down the steep slope, becoming very heavily crevassed as it does so. Most of the crevasses intersect one another, producing towers of ice called séracs that lean over and ultimately topple over.

well-defined sets, classified according to their geometry: longitudinal, marginal, transverse, splaying and *en echelon*. Sometimes, however, several sets of crevasses may intersect, creating a chaotic, totally broken-up surface, with ice towers known as **séracs** (from the French). Where a glacier flows over a pronounced step, the surface first fractures into transverse crevasses, before breaking up totally, creating chaotic reaches known as icefalls. Séracs are very unstable, and where possible icefalls are best avoided because of the danger of

Exploring a crevasse deep in Erebus Glacier Tongue, McMurdo Sound, Antarctica. The cavity was approached from the surrounding sea ice and through a tunnel in the ice cliff.

their collapsing. Many climbers have been killed by collapsing séracs, for example in the notorious Khumbu Icefall on Mount Everest.

A special type of crevasse, known as a **bergschrund** (from the German), occurs at the head of a glacier, generally where the gradient is very steep and the main body of ice pulls away from the more stable ice that adheres to the steep mountainside. Bergschrunds are irregular in form and individually may extend laterally for many hundreds of metres. Unless they are bridged by snow, a bergschrund forms a difficult barrier for mountaineers to cross on their way to the summit above.

Sometimes there is another cleft in the glacier higher still, adjacent to the rock wall; this is known as a **randkluft** (from the German). Such clefts form where a rock wall absorbs radiation and melts the adjacent ice, so they are not strictly true crevasses.

Surging glaciers

The ablation area of Matanuska Glacier, Alaska shows how crevasses in this zone begin to melt back soon after they have formed, allowing visitors to walk into them if care is taken.

In June 1993 scientists of the US Geological Survey observed that North America's largest glacier outside Greenland, the Bagley Icefield–Bering Glacier system in southern Alaska, was showing

signs of unusual activity. The glacier surface became increasingly crevassed and the terminus began to advance after many years of recession. This represented the onset of a short-lived phase of extremely fast flow known as a surge. During the course of the next 17 months, the glacier advanced approximately nine kilometres, delivered large amounts of icebergs and sediment to the ice-marginal Vitus Lake, and over-rode two islands at the peak of the nesting season for waterfowl and geese. The surface of the glacier became a maze of intersecting crevasses in response to the complex stresses acting within, and the upper reaches subsided as ice was rapidly drawn down to supply ice for the surge. Velocities reached a maximum of 88 metres per day.

In September 1994 the surge ceased, but by April 1995 a new surge

The huge Bering Glacier, viewed from a scheduled flight from Seattle to Anchorage, reveals a fine set of medial moraines which preserve evidence for several surges in the past. During the most recent surge, the glacier advanced dramatically into Vitus Lake at the bottom of the picture. Bering Glacier and all its tributaries cover 5200 square kilometres in area, and is thus one of the largest glacier systems in North America.

Glacier surges are relatively unusual phenomena, but result in dramatic changes when they occur. Here, during the 1882–83 surge of Variegated Glacier, Alaska, the surface became totally crevassed under the influence of velocities approaching a rate of 60 metres a day (2.5 metres per hour!).

was underway, compressing the winter ice on Vitus Lake into a series of accordion-like folds, whilst new crevasses and large rifts formed on the glacier surface, many of which filled with lakes. Once more, large quantities of icebergs and sediment were delivered to Vitus Lake, drastically altering its depth and shape characteristics, and more islands were over-ridden. A large increase in suspended sediment from outlet streams occurred, much of this reaching the open sea via the six-kilometre-long Seal River. The result was a sediment plume that by October 1995 extended some 100 kilometres to the west of the river outlet.

By October 1995 all visible surge activity had ceased, but there was concern that the increased discharge of icebergs carried to the ocean via Seal River would become a hazard to shipping, especially oil tankers from the port of Valdez. Bering Glacier has a history of surges, with earlier ones documented photographically in 1965–67,

Polythermal glaciers in the High-Arctic are prone to surging, but on a much longer time-frequency (perhaps more than 100 years) than those in temperate regions such as Alaska. In this 1996 image, Fridtjofbreen in Bellsund, western Spitsbergen is surging, hence the heavy crevassing. The surge triggered a rapid advance of the snout, the frontal cliff becoming heavily shattered, with fallen ice blocks being progressively overridden.

1957–60, 1940 and 1920. This history of surging is reflected in its fine suite of folded moraines.

The most closely monitored surge was that of 1982–83 by Variegated Glacier, also in Alaska. A large team of glaciologists undertook detailed studies of ice movement, ice thickness changes, ice structures, water flow through the glacier and sediment discharge. Glaciologists had in fact predicted the surge, since four similar events had been recorded earlier in the century. The scientists were able to investigate both the event as it unfolded and its aftermath. Variegated Glacier has yielded more information about surging than any other glacier to date.

During the surge, the surface of Variegated Glacier broke up totally into a maze of crevasses and chasms. The ice reached top speeds of 63 metres per day. Ice in the middle part of the glacier was displaced by about two kilometres during the 18-month-long surge, compared with only one kilometre in the entire period of 17 years preceding it. By 1986 the crevasses had melted back so that, although there remained a chaotic maze of hills and valleys of ice, it was possible by wearing crampons and taking a great deal of care and effort to walk around and examine the structures that had resulted from the surge.

Where do surging glaciers occur?

The geographical distribution of surging glaciers is peculiar and does not follow any obvious pattern. Scandinavia and the Southern

Aerial photograph of surge-type glaciers in Svalbard. Bakaninbreen (BB) is coming to the end of its surge in 1995. The surge front (SF) is a steep ramp, separating active ice from stagnant ice. The adjoining Paulabreen (PB) is a surge-type glacier in its quiescent phase. (Photograph courtesy of Norsk Polarinstitutt, Tromsø. Photograph number S90 6828).

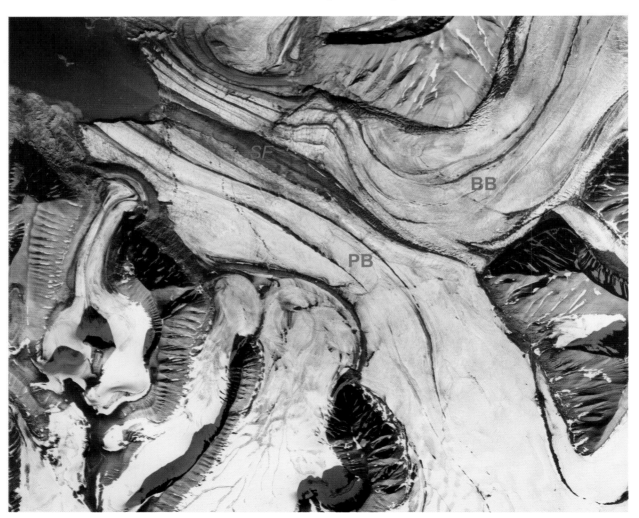

Alps of New Zealand are regarded as surge-free zones. In the European Alps historical records indicate that Vernagtferner in Austria surged, and in 2002 the Ghiacciaio del Belvedere in Italy was showing signs of surge-type behaviour. In Alaska and the Yukon surge-type glaciers are confined to the St. Elias Mountains, the Alaska Range, the Wrangell Mountains and the Chugach Mountains. Others occur in the High-Arctic Queen Elizabeth Islands. However, there are none in the more southerly parts of the coastal ranges and Rocky Mountains of Canada and the USA. Elsewhere, surge-type glaciers are found in Patagonia, Iceland, Greenland, Svalbard, several parts of the former USSR and the Karakorum Mountains. Thus, they occur in a wide range of topographic and climatic settings, but form no more than a small percentage of the total number of glaciers in any one area, except in Svalbard where at least a third of all glaciers are believed to demonstrate surge characteristics.

Some glaciers, such as the Jacobshavn Isbrae of northwestern Greenland, flow constantly at 'surging' speeds, and others, like the Fox and Franz Josef glaciers of New Zealand, occasionally advance at very fast rates. However, none of these are strictly surge-type glaciers because they lack a quiescent phase.

One of the great unresolved questions of glaciology is whether the Antarctic Ice Sheet has ever surged in the past or may do so in the future. Some glaciologists have suggested that one of the largest ice drainage basins of the ice sheet, the Lambert Glacier–Amery Ice Shelf system, might currently be building up to a surge. Others disagree, and the data available are inadequate to come to any definitive answer. The West Antarctic Ice Sheet is considered by some glaciologists to be particularly unstable because it is grounded below sea level and is held back by ice shelves that may be in danger of collapse.

A major surge of either the Lambert Glacier or the West Antarctic Ice Sheet into the Southern Ocean would have a drastic effect on the world's weather. A surge would be accompanied by break-up as the ice discharged into the sea releasing so many icebergs that the water

Some examples of ice velocities in different types of glacier

Glacier	Region	Centreline velocity (metres/year)	Comments
Lambert Glacier	East Antarctica	347	Part of largest glacier drainage system in Antarctica
Amery Ice Shelf	East Antarctica	1200	As above
Byrd Glacier	East Antarctica	760	Glacier drainage from polar plateau through the Transantarctic Mountains
Ice in West Dronning Maud Land	East Antarctica	1–15	Slow moving part of ice sheet where not channellized
Jakobshavn Isbrae	NW Greenland	4700 (max. recorded 8360)	Fastest outlet glacier from Greenland Ice Sheet
Columbia Glacier	Alaska	15 000	Accelerated receding terminus of tidewater glacier – world's fastest
White Glacier	Axel Heiberg Island, Canada	40	Cold, sliding valley glacier
Grosser Aletschgletscher	Swiss Alps	200	Fastest part of largest glacier in the Alps
Griesgletscher	Swiss Alps	40	Small valley glacier
Saskatchewan Glacier	Canadian Rockies	117	Valley glacier
Charles Rabots Bre	Okstindan, Norway	8	Steep, thin cirque glacier

and air temperatures north of the continent would be lowered. A surge would release so much grounded ice that huge volumes of water would be released, in the case of the West Antarctic Ice Sheet raising the sea level globally by several metres. Low-lying countries such as the Netherlands and Bangladesh as well as many large cities all over the world would become vulnerable to flooding. This may not be as far-fetched as it seems. There is geological evidence that the Laurentide Ice Sheet of North America surged repeatedly over the

Mid-West during the last ice age (the Wisconsinan Glaciation), about 10 800–14 000 years ago. Similarly, some geologists have proposed that an ice sheet over the Barents Shelf surged over northernmost Europe during the late Weichselian Glaciation at about the same time, before collapsing.

The nature of surges

Some surges occur at fairly regular intervals. For example, the Variegated Glacier has had seven surges in the last 100 years, the most recent in 2002. Thus, the approximate timing of the 1982–83 surge of Variegated Glacier was predicted on a known periodicity of approximately 14 years. As explained above, this enabled glaciologists to plan a major field campaign to investigate the causes of surging. Similarly, the Medvezhiy Glacier in the Russian Pamirs has a record of surging about every 10 years, and has also been intensively studied. However, the surge cycle for most glaciers is much longer. Thus, predicting when a surge event might occur is unreliable. Between **surge phases**, glaciers experience years or decades of inactivity known as the **quiescent phase** that is characterized by melting of stagnant or slow-moving ice and the development of a complex internal drainage network, including numerous large potholes.

When a surge begins, the first indication is the sudden appearance of thousands of crevasses, which heralds the onset of more vigorous flow. This break-up of the surface spreads rapidly downglacier and the maximum ice movement accelerates from perhaps only a few centimetres a day to as much as 100 metres a day as the zone of surging ice (the **surge front**) passes by. If the surge front reaches the snout, the glacier advances in dramatic fashion.

In many cases surges result in a glacier advancing several kilometres over a few months, but other surges take some years to be completed. The largest recorded surge was from an icefield in Svalbard, when the outlet glacier named Bråsvellbreen advanced 20 kilometres into the sea along a 30-kilometre-wide front some time between 1936 and 1938. Occasionally, however, the surge peters out before the snout is reached, leaving a steep ramp or bulge in the zone between the once active and inactive ice of the snout. The 1982–83 surge of Variegated Glacier, for example, ceased a kilometre from its snout.

Surges are accompanied by a tremendous release of meltwater, and the combination of severe flooding and rapid ice advance has been known to cause considerable damage down valley. During a surge ice is transferred from the higher reaches of the glacier to the snout. This results in thinning by as much as 50–100 metres in the upper part (and often leaving blocks of ice stranded on the hillsides). Conversely, thickening takes place in the lower part.

During their quiescent phases, many surge-type glaciers can be identified from the pattern of medial moraines: instead of lying in relatively straight lines more-or-less parallel to the valley sides, debris is contorted into loops. A loop develops when a tributary glacier flows into the main one while the latter is inactive, and a surge of the main glacier will carry it downwards several kilometres. Repetition of this process in several tributaries can create a complex set of contorted moraines.

Why do glaciers surge?

Several mechanisms have been invoked to explain glacier surges. At first it was thought that earthquake-generated avalanches might trigger them. More recently scientists have tried to link surges to glacier size, shape, orientation and gradient, climate, bedrock type and thermal regime. However, none of these hypotheses has furnished an adequate explanation.

In recent years two further theories have evolved from detailed observations of surge-type glaciers. Both relate to what happens at the glacier bed, one looking at the distribution of water in contact with a hard rocky bed, the other at the changing properties of soft deformable sediment. In the first case, a picture has been built up following studies during the 1982–83 surge of Variegated Glacier. Scientists have established that the upper part of the glacier (the **reservoir area**) gradually thickens over a period of time before the surge, while the lower part thins. As a result the glacier becomes progressively steeper, and the stresses at the base of the glacier in the

Iceberg Glacier on Axel Heiberg Island is a surge-type glacier, here in its quiescent state (1977), recognizable by the contorted moraines and pitted surface resulting from streams in stagnant ice disappearing down moulins or potholes.

reservoir area increase. Thus, the subglacial meltwater channels close more easily. Eventually, the channels are squeezed shut and the water is forced to spread out as a film across the entire glacier bed. This drastically lowers the friction by separating the ice from bedrock, and the glacier begins to slide much faster. Rapid sliding usually starts in the reservoir area, but as the fast-moving ice impinges against the slower-moving ice down-valley, very high stresses occur and the surge begins to spread. The resulting **surge front** propagates down-glacier as a **kinematic wave** that moves faster than the ice itself, rather like the fast and slow pulses of bunched vehicles in high-density road traffic.

After the surge the glacier is less steep. Therefore, the stresses decline and once again allow the opening of subglacial channels, at which point the glacier ceases to slide and drops back on its bed. The newly formed channels enable the subglacial reservoir of water to empty, providing the final dramatic flood. At this point in the cycle

Relative importance of the three different types of flow in individual glaciers

Glacier	Region	Internal deformation (%)	Basal sliding (%)	Subglacial deformation (%)
Taylor	Antarctica	40	60	0
Blue	Washington, USA	10	90	0
Breiðamerkurjökull	Iceland	0	12	88
Trapridge	Yukon, Canada	12	50	38
Urumqui No 1	China	37	3	60
Storglaciären	Sweden	23–43	57–77	0
Variegated (non-surge)	Alaska	100	0	0
Variegated (surge)	Alaska	5	95	0

After Knight, P. J. *Glaciers*, Cheltenham: Stanley Thorne (Publishers) Ltd., 1999.

the ice velocity of Variegated Glacier dropped from 30 to barely 3 metres a day within only 24 hours, but its final flood transported huge amounts of sediment, burying numerous shrubby trees in front of the glacier.

An alternative, and perhaps complementary, explanation has arisen from studies of the Trapridge Glacier in Yukon Territory. Being a polythermal glacier, it has a more complex thermal structure than Variegated Glacier, since it is cold in part. It is believed that the glacier rests on a bed of soft sediment, mainly till, rather than bedrock. Drainage is by way of channels incised into the sediment during the quiescent phase. As ice builds up in the reservoir area, stresses at the base increase and the channels are no longer able to remain open. Water is then forced to flow through the sediment, which consequently is weakened and therefore prone to rapid deformation. A cushion of deforming sediment is formed in which rapid acceleration of the ice becomes possible. Trapridge Glacier has been progressively building up to a surge over several decades, but tantalizingly had still not surged at the time of writing.

From the above investigations, it is clear that the common factor that triggers surging is changing conditions at the glacier bed, but

Glaciers with more than two recorded surges

Glacier	Country	Year of surge				
Bruarjökull	Iceland	1625	1720	1810	1890	1963
Carroll	Alaska	1919	1943	1966		
Kolka	USSR	1834	1902	1969		
Medvezhiy	USSR	1937	1951	1963	1973	
Nevado Plomo*	Argentina	1788	1934	1985		
Variegated+	Alaska	1906	1947	1964	1983	1994
Vernagtferner#	Austria	1600	1678	1773	1845	

Notes:

* Two or more additional surges may have occurred between 1788 and 1934. The surge of 1788 was not directly observed but can be inferred from a lake outburst.

\+ There was probably also a surge around 1926 according to anecdotal evidence.

\# The surges prior to 1845 are inferred from lake outbursts.

other mechanisms should not be ruled out. There is still much to be learned about why surges take place, but understanding them is crucial if we are to be able to differentiate between major climatically induced advances of the major ice sheets and surge-related advances.

Impact of surges on human activity

Although most surges take place in remote areas, we must learn to predict surges where people or installations are endangered in order to be able to take evasive action. In Alaska the Black Rapids Glacier is being monitored because a surge might threaten the Alaska Pipeline and the Richardson Highway. There is good cause for concern, as the glacier has already issued one warning, for during the winter of 1936–37 the occupants of a lodge on the Richardson Highway were startled to see the three-kilometre-wide front of the glacier bearing down upon them. The normally smooth gentle snout

False-colour Landsat image of the huge piedmont lobe of Malaspina Glacier, southern Alaska. The glacier is represented by shades of white and blue. The spectacular fold structures are the result of a combination of surging and compression in the direction of ice flow. The red colour represents vegetation, some of which is growing around the margins of the glacier. (Photograph courtesy of NASA; image taken 31 August 2000 by Landsat 7.)

had been transformed into a heavily crevassed ice cliff 100 metres high, and the ice was advancing at rates of up to 66 metres a day. If the surge had continued, it would have dammed a major river, severed the highway and demolished the lodge. Fortunately for the occupants of the lodge and the people of Fairbanks, who relied on the road for links with the outside world, the glacier stopped a short distance from the road.

In the European Alps, a region mostly devoid of surge-type glaciers, a remarkable advance of the Ghiacciaio del Belvedere on the east

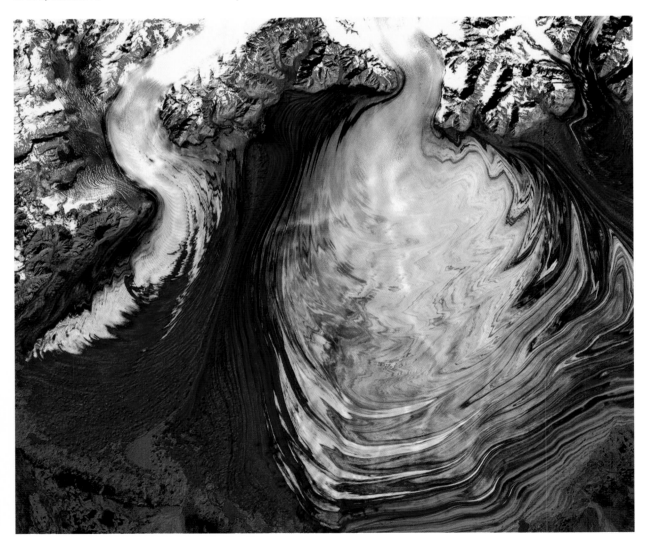

face of Monte Rosa in Italy has been taking place recently. A surge-like acceleration of ice flow, accompanied by heavy crevassing, began in the winter of 2001–02, and was continuing at the time of writing. The serious hazard implications of this event are described in Chapter 13.

All things considered, therefore, glacier surges are perhaps the most exciting and intriguing of glaciological phenomena. Whilst most surges occur far removed from civilization, where they do have an impact directly on humans, close monitoring is essential.

6 Nature's conveyor belt

Few natural processes on the Earth's surface can match glaciers in transporting debris long distances away from their source. Indeed, one of the most immediately obvious features of mountain glaciers is the amount of rubble and the huge blocks of rock that litter their surface. Commonly, the entire ablation area of a mountain glacier is completely debris-covered, although the melting of ice beneath the debris layer creates a very uneven and unstable topography that is arduous to walk over. Thus, a glacier may be considered as a sort of conveyor belt for rock debris, transporting material from all points along its length towards the snout. Typically, this material is carried on the surface (where it is referred to as **supraglacial debris**), or near its bed in basal ice (where it is described as **basal debris**). Large amounts of debris may also be ingested from both the surface and bed as **englacial debris**, giving the glacier a very dirty appearance. In addition to debris carried within the ice, debris is also transported within the deforming bed zone.

Surface debris

For a glacier to be laden with supraglacial debris, it normally requires rocks to be exposed above its flanks. Thus, ice sheets and ice caps, which submerge mountains, carry almost no supraglacial debris, whereas the lower reaches of mountain valley glaciers carry many boulders. The main sources of this supraglacial debris are rock-fall from frost shattering, aided by the unstable nature of the hillsides over-steepened by the glacier. Most rock-falls are small, although occasionally entire landslides cover broad expanses of ice, especially in regions prone to earthquakes, such as Alaska, the Andes, the Himalaya or the Southern Alps of New Zealand.

Huge amounts of debris are transported and deposited by valley glaciers. This photograph shows Breithorngletscher (right) and Schwarzgletscher (centre) flowing off Breithorn (4164 metres) near Zermatt in Switzerland to join Gornergletscher flowing from left to right. Note how lateral moraines become medial moraines where glaciers flow together.

Supraglacial (or surface) debris is normally the result of rock-falls, but occasionally larger landslides occur, as in the accumulation area of Glacier Pré du Bar in the Mont Blanc area, French Alps.

A well-publicized large-scale rock-fall was the collapse of the summit of Mount Cook (Aoraki), New Zealand's highest mountain. Just after midnight on 14 December 1991, climbers, in preparing for the ascent of the mountain at Plateau Hut, heard a loud rumble and felt their accommodation shake. Looking out into the darkness, they saw bright orange flashes before the hut was enveloped in a cloud of dust. Deciding that it was unwise to climb the mountain, they found next morning that falling rock and ice had come within 300 metres of the hut, and that the entire glacier below the summit was obliterated. In fact, the rock-fall swept across the ice terrace of the Grand Plateau, down the Hochstetter Icefall, fanned out across the two-kilometre-wide Tasman Glacier below, and up the other side of the glacier for 70 metres – a 2.7-kilometre vertical fall and a horizontal distance of seven and a half kilometres. New Zealand Government scientists subsequently calculated that 29 million cubic metres of

The parallel 'tramlines' on the Grosser Aletschgletscher, Berner Oberland, Switzerland are medial moraines, each formed where two ice streams combine. The arcuate structure between the moraines is foliation, a layered deformation structure in the ice. The three peaks in the background, from left to right, are Jungfrau (4158 metres), Mönch (4099 metres) and Eiger (3970 metres).

snow, ice and rock had been involved, largely derived from the eastern face of Mount Cook, but taking with it the summit, thereby reducing the mountain's height by 10 metres to 3754 metres. Although there were no fatalities or damage, the event made the news throughout the world, and was even recorded by a seismograph in Wellington, 500 kilometres away. Although Mount Cook lies near a fault, the underlying cause appears to have been the undermining of highly fractured rock by freeze–thaw processes.

Earthquakes frequently cause spectacular rock-falls onto glaciers in Alaska. An earthquake in 1964 led to the covering of the lower reaches of Sherman Glacier with debris, slowing ablation so much that the glacier started to advance. More recently, the Denali Fault earthquake of 3 November 2002, which measured 7.9 on the Richter

Opposite. Supraglacial debris may be more evenly scattered over the glacier surface if rock-falls are intermittent. This shows a view down the Gornergletscher, looking towards the Matterhorn (4477 metres).

Where the ice slows down and becomes crevassed, medial moraines merge into one another, as on Hubbard Glacier, Alaska. Despite now merging into a continuous cover of debris, the individual moraines may be distinguished on the basis of the different-coloured rock types.

scale, resulted in spectacular changes to the landscape. Of these changes, huge rockslides that crossed the Black Rapids Glacier are especially impressive. These rockslides are so large that reduced melting in the ablation area may lead to an advance like that of Sherman Glacier.

The size of the rocks carried by the glacier depends on the rock type: cliffs of resistant, unbedded rocks like granite produce huge blocks, often the size of small houses, but softer rocks like shale or limestone invariably break down into small boulders. Generally, supraglacial debris contains relatively small amounts of sand and finer material.

Most debris falling down mountainsides becomes caught up with the ice as it slides past the valley sides. The lines of debris on the surface at the edge of the glacier and the ridges of debris left behind as the glacier recedes are known as **lateral moraines**. Where

Opposite. Supraglacial streams often wash out the finer material (sand and pebbles) from the moraine cover and transport it down-glacier. If it collects in hollows and the stream is rerouted, the debris protects the ice from subsequent ablation, so that ultimately it stands proud of the ice surface as a dirt cone. This three-metre-high cone on Glacier de Tsijiore Nouve, below Pigne d'Arolla (3796 metres), Switzerland, is typical in comprising a thin cover of debris over ice.

Isolated blocks on the glacier surface often protect the ice from melting, especially when solar radiation is strong, forming glacier tables. This example is on Vadret Pers, southeastern Switzerland.

two streams of ice join, the two lateral moraines combine to form a single **medial moraine**, which appears as a line of debris extending towards the snout, down the middle of the glacier. Commonly, moraines comprise a variety of rock types that reflect the different source areas. When ablation near the snout brings medial moraines of different composition into close proximity, a glacier may take on a multicoloured striped appearance, as in the justly named Variegated Glacier in Alaska or the Unteraargletscher in the Alps.

Dark debris on the surface of a glacier absorbs the Sun's radiation better than the surrounding ice. The debris cover becomes relatively warm, and increases the melting of the ice, particularly if the debris is no more than a few centimetres thick, with the result that the moraines occupy depressions. In contrast, a thick continuous cover of debris normally slows down ablation, so the debris stands proud above the general glacier surface as a distinct ridge.

Isolated boulders also protect the ice from melting, so they may end up sitting on perches of ice. These **glacier tables** often tilt towards the Sun as time goes by, until they slide off, and the process is repeated.

Streams on the surface may recirculate much of the finer debris, some of which collects in hollows. Once a stream course has been abandoned, the surrounding surface continues to ablate. However, the debris in the depression may slow down the melting of the underlying ice, so that sand or gravel eventually rests on small rises. As the surrounding ice carries on melting these rises can become **dirt cones**. If you hack at them with an ice axe you find that they consist simply of a veneer of debris a few centimetres thick covering a cone of ice.

The tongue of Chola Glacier in the Khumbu Himal of Nepal is completely debris-mantled. The thick debris-cover has prevented the glacier in the current phase of climatic warming from receding. The steep flanks of the glacier are bounded by lateral moraines which merge imperceptibly into glacier ice beneath the debris cover.

The proportion of debris cover increases towards the snout in most valley glaciers. Uneven melting of ice beneath the debris cover and the action of streams on slow-moving or stagnant ice create an irregular surface of sharply defined hills and valleys with a relief of several metres. The debris is often unstable and, although not technically difficult to walk on, it can nevertheless be hazardous. During the ablation season, debris is continuously falling down the ice slopes, many of which have only a thin cover of debris. Such terrain, with its deeply incised channels, pools and englacial streams, is referred to as **glacier karst** by analogy with similarly eroded areas of limestone.

If, as a result of climatic warming, the glacier wastes away, supraglacial debris is lowered on to the bed beneath, or on to debris

Many glaciers in high mountain regions are completely mantled by debris in their lower reaches. This view shows the uneven surface of loose debris and supraglacial ponds on the surface of Khumbu Glacier, Nepal.

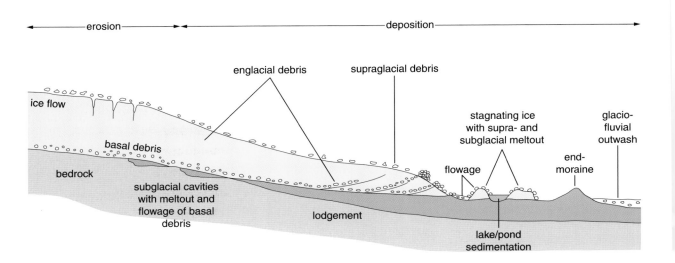

Figure 6.1 A longitudinal profile through the lower reaches of a receding land-based glacier, illustrating how debris is transported and deposited.

being deposited from the base of the glacier. Such deposits are known as **supraglacial meltout tills** (till being the term for the poorly sorted deposits released directly from ice). The deposits are irregularly spread over the ground and are constantly reworked by sliding and slumping, and by streams, until sufficient vegetation grows to stabilize them.

Debris-covered glaciers

In extreme cases, the supply of rock-fall material may be sufficient to completely cover a glacier tongue. Such glaciers are visually striking, although hardly beautiful. Debris-covered glaciers are normally flanked by high mountains and the main source of debris accumulation is via ice/rock avalanches. Avalanche-fed glaciers are dominant in the tropical Andes and parts of the Himalaya.

The effect is to create a very uneven surface with a relief of tens of metres, steep ice faces and ponds. Walking across such glaciers requires extreme care, as even the largest blocks can move without warning.

Debris-covered glaciers have taken on unusual importance as climate warms and they begin to recede. They do not behave like debris-free glaciers. Firstly, the debris cover allows the glaciers to survive at lower elevations than if they were debris-free. Secondly, as shown by Japanese glaciologists on the Khumbu Glacier in Nepal, the ablation rate *decreases* down glacier because the increasingly thick debris mantle reduces the melting of the underlying ice. As the glaciers recede, the middle reaches melt the fastest, resulting in the development of a hollow. If meltwater cannot readily escape through the bounding moraines, a lake may form by the coalescence of small ponds. Ultimately, a large lake may develop behind a terminal moraine that is composed of loose unconsolidated glacial debris and possibly a decaying core of stagnant ice. In Nepal and Peru such moraines may exceed 100 metres in height, so the potential for a major catastrophe following a moraine-dam failure is great. The formation of moraine-dammed lakes is described in the following chapter, and examples of associated disasters are given in Chapter 13.

A considerable amount of debris can be transported at the base of a glacier. Up to several metres of debris and water-ice may freeze to the bed, producing layered dirty ice, the basal ice layer. Here at Taylor Glacier in the Dry Valleys of Antarctica, not only can the basal ice layer be seen, but also the lighter coloured 'till' deposited from it.

The Chugach Mountains (far left) and the St. Elias Mountains (centre and right) are among the largest highland icefields in Alaska. This satellite image shows approximately 300 kilometres of the Gulf of Alaska coastline where many glaciers flow into fjords or approach the open sea. These glaciers deliver large quantities of suspended sediment that appear as light grey or light blue plumes along the coastline, in contrast to the dark blue of the sediment-free ocean waters. The largest glaciers, featured elsewhere in this book, include the long complex valley glacier, the Bering (left); the piedmont glacier, the Malaspina (centre); and the Hubbard entering Yakutat Bay and almost sealing off Russell Fiord (middle right). Photograph courtesy of NASA; image taken 22 August 2003; source: http://rapidfire. sci.gsfc.nasa.gov/gallery/?2003234-0822/Alaska.A2003234.2105.1km.jpg.

Debris transported along the glacier bed

Debris at the base of a glacier (**basal debris**) is very different in character from supraglacial debris, since it is continually being modified as the ice moves downhill. Before a glacier can pick up debris the ground may have been deep-frozen and then thawed, processes that loosen blocks of rock, especially along bedding planes. The blocks and other loose debris then freeze onto the base of the glacier. Further debris is collected as a result of small-scale changes in pressure around bedrock bumps, which melt and refreeze the ice: as the ice passes over a rise, the pressure rises, thereby lowering the melting point; then on the downstream side the pressure lowers and the water refreezes, producing **regelation ice**.

Once detached from the bed, the blocks of rock become powerful tools for eroding the bedrock further. Held firmly in the ice, the rock fragments groove and scratch (striate) the bedrock, and themselves have their sharper corners broken off. They also rotate within the ice because the velocity of the glacier increases upwards, so new parts of the blocks are constantly being exposed to the bedrock. Thus a detached block that started angular becomes progressively more rounded and acquires striations, some stones becoming quite round (although not as round as those carried by streams). Most basal debris is picked up beneath the accumulation area where the ice is more dynamic, or along the valley sides. It is in these areas that the sandpaper effect of the glacier sole is greatest and most erosion of bedrock occurs. In the ablation area, where the glacier is slowing down, more of the debris load is deposited as **basal till**. These processes of grinding and crushing at the glacier bed also generate finer material, including clay and silt – the **rock flour** that gives the characteristic milky appearance to glacial streams.

The processes of deposition are complex, but two main types of basal till result from it. At first, with the ice deforming rapidly, melting near its bed and sliding rapidly, debris is actively plastered on to the bed to give a **lodgement till**. The lodgement process becomes less effective towards the snout and direct melting out of debris from the ice takes place, producing a **meltout till**. Both types of till contain a mixture of particles ranging in size from clay to boulders.

Sheets of till cover much of North America and Europe, where they provide fertile mineral-rich soil for agriculture. However, tills sometimes generate problems for the construction industry because, although commonly providing a firm base, when wet they can deform easily, and on slopes they can be subject to land-slippage.

Debris inside a glacier

A certain amount of debris becomes incorporated in the interior of a glacier as **englacial debris**. Surface debris from rock-falls in the accumulation area becomes buried by snow, and other debris falls down crevasses, although in both cases it remains below the surface until released by ablation. Debris may also be found in an englacial position along thrusts, which are faults extending upwards at a low angle from the bed. Some of this debris may even reach the surface, especially near the snout, but its basal origin can be recognized from the part-rounded and striated nature of the stones.

There are differences in the amount of debris carried according to the temperature of the glacier. Polythermal glaciers tend to have a very thick basal debris load, perhaps representing 50 per cent of the total ice thickness towards the snout. This is probably the result of a combination of freezing of debris to the base and folding in the basal part of the glacier, repeating particular debris layers. In contrast, the amount of debris carried on the surface is small. The reverse is true for temperate glaciers. These commonly have an extensive debris cover, whereas the basal debris layer is normally less than a few metres thick. Many glaciologists have argued that cold glaciers, which are frozen to their beds, tend to be inactive in terms of their effect on the landscape, but rock fragments embedded in the ice may erode the bed to a limited extent, despite the lack of sliding.

Kronebreen in northwest Spitsbergen produces large numbers of icebergs. Many are rich in basal glacial debris, and float across the waters of Kongsfjorden, slowly releasing their load onto the sea bed as dropstones. The dark grey iceberg, so dirty it could be mistaken for an island, dwarfs the person standing on the shore beyond.

Furthermore, there is some evidence that the large ice sheets of North America and Eurasia ripped up large blocks of rock from the bed even when the ice was frozen.

Debris in glaciers also includes material that has been carried by the wind, and most glaciers have minor amounts of wind-blown silt derived from glacial rock flour deposited by streams beyond the confines of the glacier. Sometimes the debris originates in storms far distant; for example, every couple of years or so, yellow Saharan dust is blown onto Alpine glaciers, staining their surface. In other places, notably Iceland, the Andes and Alaska, volcanic eruptions have thrown significant amounts of ash on to a glacier. If the time of the eruption is known, the ash layer can tell glaciologists a lot about glacier mass balance and dynamics.

Other types of debris in ice are volumetrically insignificant, but are important in providing clues about environmental change, such as traces of industrial pollution or forest fires, and radioactive fall-out from nuclear explosions. Studies of the acidity of snow and ice have also provided much information about past volcanic eruptions, even when the actual dust and ash are barely visible to the naked eye.

7 Ice and water

Glacial meltwater plays an important role in the landscapes around glaciers, and in some cases in the lives of the people who live near them. For example, in Switzerland, up to 15 metres of ice may melt vertically from the lower lying parts of the Grosser Aletschgletscher every summer and help generate the hydroelectric power used by the totally electrified Swiss railway system. In the arid regions of northwestern China and in Argentina glacial meltwater irrigates desert land, ensuring the survival of many thousands of people. In the Cordillera Blanca of Peru, complex irrigation networks connected to glacial meltwater streams supply varied crops on steep mountainsides during the dry season. Major cities, such as Lima and La Paz, rely heavily on glacial meltwater for sustaining millions of people.

Meltwater also plays a significant part in the development of glacial landscapes. In high polar regions, during the short summer season, snow and ice melt combine with rainwater to provide a noisy, tinkling, gurgling, rushing or roaring background to any activity near glaciers. Glaciers everywhere generate their own stream systems, either on their surface or within and below the ice, in a similar manner to streams in limestone regions. During the peak period of melting in early summer the stream that emerges at the snout of a glacier is often a spectacular torrent, frequently flooding the valley floor below. Yet in winter, discharge is reduced to a mere trickle and in many parts of the world water is locked solid as ice for as many as nine months of the year. These extremes between summer and winter provide a fascinating range of meltwater features on and around glaciers.

Surface ('supraglacial') meltstream on Austre Lovénbreen, a small valley glacier in northwest Spitsbergen. In summer, streams on Arctic glaciers become more incised and can be difficult to cross. Streams often retain the same channel year-after-year.

Factors affecting melting

The principal influence on melting is, of course, air temperature, although even on bright sunny days with sub-zero temperatures, melting may be significant because of intense radiation. Practically all glacierized regions except Antarctica experience daytime summer temperatures several degrees above freezing. During sunny weather the Sun's radiation generates large volumes of meltwater, but at night there is usually an almost total freeze-up. Consequently, meltwater discharge fluctuates on both a daily and a seasonal basis. Extremes of daily discharge become more pronounced with distance from the poles. A practical aspect of these daily variations is that, whereas a meltwater stream may be easily crossed in the morning, by mid-afternoon it may be impassable, and it is therefore quite easy to get stranded on the wrong side. In cloudy weather daily extremes are not so pronounced and, although daytime meltwater production may be reduced, melting will continue at a slightly lower level during the night.

Another agent promoting melting is geothermal heat – the heat from the Earth's interior. In its most extreme form, such as in volcanic regions, geothermal heat can melt large volumes of ice and create subglacial lakes, but normally it only produces limited melting at the glacier bed. Nevertheless, such basal melting is a significant process beneath many otherwise cold glaciers, such as those in the High Arctic or Antarctica. Although the bulk of the ice may be at sub-zero temperatures, and in the absence of geothermal heat would reflect the mean annual air temperature at the site measured, geothermal heat causes the temperature to increase with depth in the glacier. Indeed, if the ice is thick enough, melting occurs at the base. In such a case, the glacier behaves like an insulating blanket over the land: quite moderate temperatures are found under the glacier, while the ground around the glacier may be permanently frozen to a depth of several hundred metres. Frictional heat, generated as a glacier slides over its bed, or as a result of internal deformation of ice crystals gliding over one another, also can generate meltwater.

Lakes at the margins of glaciers are ephemeral features and can drain suddenly. These two photographs of Pastaruri, an icefield in the Cordillera Blanca, Peru, show the site of a drained lake. The smooth semi-circular tunnel was the route taken beneath the ice when the lake drained suddenly.

Snow swamps

The pattern of melting differs between cold and temperate glaciers. Cold glaciers commonly do not have a well-defined firn line in summer, and it is common for large areas of the winter snow pack to become saturated with water in early summer because the cold ice prevents a good internal drainage network from developing. These areas of saturated snow are known as snow swamps. They are treacherous because their surface may appear like any area of dry snow, but beneath may lie metres of slush or even streams. They are often unstable, and it is common for any disturbance, such as a person walking across them, to trigger a slush avalanche, or **slush**

On alpine glaciers, such as Vadrec del Forno, streams commonly exploit structures in the ice. In the longitudinal foliation shown here, dark ice layers absorb more solar radiation than light layers, so forming grooves that then provide ready-made channels for meltwater.

flow. Usually such flows are small, but occasionally large areas of the snow pack become active and whole areas of a glacier are swept clear, exposing bare glacier ice. As the saturated snow comes to rest and the water escapes, the slush packs into a very hard mass of dense snow. If a saturated snow pack remains in place, the first indications of meltwater will be vague channels that appear deceptively small. Anyone unfortunate enough to fall in may find it extremely difficult to get out.

On temperate glaciers the firn line is clearly demarcated and normally only a narrow zone of saturated snow develops. Stream courses are also well defined from an early stage in the melt season, because it is easier for meltwater to drain downwards to the bed than it is on cold or polythermal glaciers.

Stream channels that become reoccupied each year on polythermal glaciers evolve slowly through time independently of ice structures. On Vibeke Gletscher in East Greenland, an exceptionally fine set of meanders has formed, and are here viewed from a helicopter.

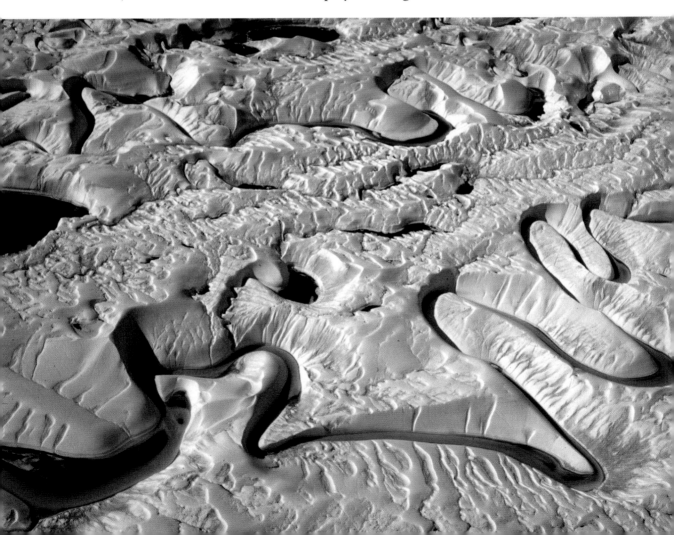

The glacier drainage system

In temperate glaciers, in which ice is at the melting point throughout, a system of veins develops at the boundaries between ice crystals. This attribute allows glacier ice to be permeable at the microscopic scale, and even develop a **water table** below the surface. However, most water migrates through a glacier via a network of **channels** and **conduits**. The development of meltwater channels on the surface of a glacier depends on the rate of melting, the rate of deformation of the ice, the extent of crevasses and the pattern of other structures such as foliation, and ice temperature. Surface channel systems develop best on stagnant and on polythermal glaciers, but will not appear at all on those with a large number of crevasses. The channels themselves range in size from tiny **rills** a few centimetres across to canyons several metres deep and wide, and hundreds or so metres long that form an impassable obstacle. On

If the surface glacier ice is cold, i.e. below the melting point, water is unable to drain away freely, so ponding takes place, as here on Thompson Glacier, Axel Heiberg Island, Canadian Arctic.

flat, crevasse-free glaciers, the streams may form into a dendritic pattern – like branches of a tree joining to form the main stem. Alternatively, they may form tight meandering patterns, with deeply incised channels marked by undercut walls on the outside of bends.

Glacier drainage is commonly influenced by the distribution of surface debris and ice structures such as foliation. Medial moraines often stand out as ridges within a larger longitudinal depression. Streams may also flow parallel to foliation, since different ice types melt at differing rates, forming a characteristic ridge-and-furrow topography. Glacier structure also controls drainage in the vertical dimension. Some ice structures, especially the traces of former or developing crevasses, act as planes of weakness that are exploited by meltwater to form a glacier mill or **moulin**, similar to a pothole in limestone country. These range in diameter from a metre or less to as much as 10 metres. It is through them that much of a glacier's

One of the rare polythermal glaciers in the Alps is Gornergletscher. Those parts of the glacier that originate above 4000 m are cold. The pond is one of several that are found in the cold ice that has reached the glacier tongue. The Matterhorn is in the background.

surface meltwater reaches the bed, or at least an internal drainage network, and a view from the edge of a moulin down which a large meltwater stream falls presents an imposing sight.

Glacier surface streams are often difficult to cross. Although narrow, the ice-cold water and the slippery sides and bottom of a channel make jumping hazardous, and one may have to walk many kilometres to get round them. Unlike normal streams, it is often better to walk *downstream* to cross one as there is a good chance that the water will disappear down a moulin.

Small pools of standing water may develop on the flatter parts of a glacier, especially where dark patches of debris or dust absorb more solar radiation and melt down into the ice. The smallest examples, known as **cryoconite holes**, are cylindrical tubes a few centimetres across, but maybe tens of centimetres deep. The impression

In temperate glaciers, streams commonly emerge from the snout at a single outlet known as a *glacier portal*. The water is commonly laden with suspended sediment and the roof of the portal prone to collapse, sending 'ice boulders' down-river. This view is of the snout of Fox Glacier in the Southern Alps of New Zealand.

is that the debris in the bottom has bored its way down into the ice. In sunny weather cryoconite holes develop in large numbers and occasionally give the surface of a glacier a honeycombed or pitted appearance. The holes merge to form larger ponds and, in extreme cases, lakes tens of metres across. Such features normally have near-vertical sides, and their levels may fluctuate between day and night if they are linked to the internal drainage system of the glacier. Large lakes seem to develop mainly on cold or polythermal glaciers.

Little is known about the internal drainage systems of glaciers. Nevertheless, through the use of dye tracers it has been possible to monitor how fast water goes through them. For example, on the north Norwegian glacier Austre Okstindbreen typical rates of

Ice-cored moraines at the terminus of a glacier are associated with the development of numerous ponds and lakes. Here at Imja Glacier, Khumbu Himal, Nepal, runoff from the glacier passes through one such lake. All the loose debris in this picture is underlain by dead glacier ice, so the topography is constantly changing as it slowly melts.

Streams emanating from cold glaciers do so only from the glacier surface. The water is generally clean. This view shows the start of Onyx River, the longest in Antarctica, draining into an inland lake. Several supraglacial meltstreams run off the surface of Wright Lower Glacier, Dry Valleys area, in the background and cut through a pile of wind-blown and fluvial sands, before spreading out in braided fashion.

Supraglacial streams tend to exploit structural weaknesses in the ice and work their way towards the bed. The streams often plummet to the depths of a glacier via vertical shafts called *moulins*. This view shows the well-known site known as Les Moulins on the Mer de Glace, France that typifies this feature.

water throughput were recorded as 0.6 to 0.8 metres per second (22–29 kilometres per hour) over distances of 500–1000 metres. However, peak speeds of 1.8 metres per second (65 kilometres per hour) have also been documented. Studies on the Haut Glacier d'Arolla in the Swiss Alps show that the basal drainage system opens up as the melt season progresses, as recorded by the rate at which dyed streams pass through the system. In winter internal deformation of the ice causes the channels to close up again. Thus the efficiency of drainage networks is greatest late in the summer season.

Temperate glaciers have very different internal drainage systems from polythermal ones. In the former, most water reaches the bed well before the snout, and emerges from a single glacier portal. In contrast, in polythermal glaciers most meltwater migrates towards the margins because it cannot easily penetrate ice with sub-zero temperatures. At polythermal glaciers meltwater cuts deep channels between the hillside and the ice. Because water cannot reach the snout if the ice is frozen to the bed, a polythermal glacier does not have a glacier portal; instead, two marginal meltwater streams may flow from the glacier independently, only combining some distance from the glacier. The organization of meltwater streams from a poly-

Sediment-laden meltwater is a feature common to most glaciers. It is derived as a result of the glacier grinding down the bed. The fine components are of clay and silt size, and remain in suspension for considerable distances down-stream. Here milky brown water emerges from the lake in front of Tasman Glacier in the Southern Alps of New Zealand through a breach in the terminal moraine. The dark object partially blocking the lake outlet is a dirty iceberg.

thermal glacier means that they are more of an obstacle to gaining access to the glacier surface than is the single central stream from a temperate glacier.

Glacial lakes

Glaciers store water in a variety of situations in the form of ice-dammed, proglacial and subglacial lakes. **Ice-dammed lakes** occur where a stream from a side valley meets the glacier, where the glacier passes an embayment in the valley side, or at the confluence of two glaciers. Ice-dammed lakes are associated mainly with cold or poly-thermal glaciers, as water flow within and beneath them is impeded. However, lakes a few kilometres across are also known from some temperate glaciers. Lakes of this type commonly fill up during the melt season until the head of water is sufficient to allow escape beneath the ice, causing a flood. One example is Gornersee, formed beneath Switzerland's highest mountain, Monte Rosa, where the Gornergletscher and Grenzgletscher combine. This lake discharges in mid-summer via a ten-metre-high ice tunnel. Once this event caused floods downstream in villages such as Zermatt and Randa. However, the problem is now managed by capturing water for hydroelectric power and reinforcing the stream channels through the villages. Such floods are commonly referred to by the Icelandic term **jökulhlaup**, although strictly this term should be restricted to events caused by subglacial volcanic eruptions.

Some glaciers have rivers that flow along their flanks or across the terminal face, and are also prone to cliff collapse. Here, three onlookers watch as part of the terminal cliff of Childs Glacier, undercut by the fast-flowing Copper River in southeast Alaska, collapses.

Proglacial lakes form in two situations – in low relief areas in contact with the ice, or behind moraine ramparts in high mountain regions during glacier recession. Although the former are usually benign, the latter have the potential to cause major catastrophes. **Moraine-dammed lakes** form where debris-covered glaciers recede back from their terminal moraines, especially those that were formed during the Little Ice Age of around AD 1750 to AD 1850 in the Andes and Himalaya. As the lakes grow and the moraines subside by slow melting of buried ice, the potential for a major flood increases.

Moraine-dam failures and the resulting outburst floods have caused considerable loss of life. The challenge to glaciologists and engineers therefore is to remediate such lakes by artificial lowering or siphoning, as explained in Chapter 13.

Glaciers commonly trap water along their flanks, forming ice-dammed lakes. The water may be produced by another glacier, as here where a side valley glacier calves into Astro Lake, dammed by the much larger Thompson Glacier flowing from left to right at the bottom of the picture. These glaciers are on Axel Heiberg Island in the Canadian High-Arctic.

Two views of Between Lake, an ephemeral lake formed where the Thompson and White Glaciers on Axel Heiberg Island meet. The first picture shows the lake almost full; the second after it has drained catastrophically. This filling and emptying takes place once every year, the outlet closing up over the winter as a result of ice-creep.

Overleaf. Because glaciers often excavate deeply below the general lie of the land, when they recede they are replaced by large proglacial lakes, such as Jökulsarlon in front of Breiðamerkerjökull in southern Iceland. The deeper the lake becomes, the bigger the icebergs that are produced. Here, tourists have a good view of the spectacle, even when low cloud and fog obscure the wider view.

Icelandic jökulhlaups

Other large ice-contact lakes are those that form subglacially as a result of geothermal activity. This type of lake is common in volcanic regions, notably Iceland. When subglacial volcanoes erupt, a large amount of meltwater is produced. Icelanders use the term **jökulhlaup** to describe the resulting floods that are so typical of the area around Vatnajökull and Mýrdalsjökull in southern Iceland. Such floods occur on an almost unimaginable scale and the infrastructure in the region, particularly the main road around the island, is unable to cope when these events take place. The large flood plains, called **sandar** (**sandur** singular), are subject to major modification during jökulhlaups; hence little construction takes place in these areas, so the impact on humans is minimized. The largest jökulhlaups have been produced by the eruption of the subglacial volcano Katla beneath Mýrdalsjökull, with peak discharges estimated to be in the range of 100000–300000 cubic metres per second. The earliest recorded event at Katla was in 1625 and the most recent in 1918. During the latter, discharge peaked at about 200000 cubic metres per second, which equates to that of the Amazon River. In 2002 seismographs recorded once again earthquakes under the ice cap, perhaps heralding a new volcanic eruption. Scientists monitor Katla very closely, in order to predict a future jökulhlaup, and take precautionary measures against flood damage.

The best-studied jökulhlaup was that resulting from the 1996 eruption beneath Vatnajökull. The eruption itself is described in Chapter 9. The volcanic eruption began on 30 September, and meltwater travelled subglacially into the subglacial caldera lake of Grímsvötn. When the lake level reached a critical level on 5 November it burst out beneath the outlet glacier Skeiðarárjökull. The flood reached a peak discharge of 40000–50000 cubic metres per second within 15 hours of starting, and 3.2 cubic kilometres of water drained from the lake within 40 hours. This discharge is second only to that of the Amazon, and 20 times that of the River Rhine where it enters the sea. The effects of the flood were spectac-

In areas of flat topography, 'proglacial lakes' commonly form around the snout of a glacier. Small icebergs may calve into such lakes as here at Sheridan Glacier, near Cordova, Alaska.

ular. Water burst through the snout of the glacier in zones of structural weakness and large-scale collapse and erosion of the glacier margin took place. Large ice blocks, many bigger than a bus and weighing up to a thousand tonnes, were transported downstream and large-scale changes to the river channels took place, burying many of the blocks. Flooding affected an area of 750 square kilometres. Deposition of sand and gravel locally extended the coastline 800 metres, with seven square kilometres of new land being created. This event caused US $15 million's worth of damage to roads and bridges, a large sum of money for a country with a population of only 270 000. However, close monitoring of the eruption and growth of Grímsvötn ensured that there were no human casualties.

Implications of glacial meltwater for humans

Glacial meltwater is an important component of the glacial system. It is responsible for much of the erosion beneath glaciers, and also for the production of a wide range of depositional features (Chapter 10). Such features are not only an important part of the appreciation of glacial landscape, but meltwater deposits are valuable as sand and gravel resources. Meltwater yields many other benefits to humans, such as water for irrigation or hydroelectric power generation (Chapter 12). However, it can also be responsible for devastating floods that in the past have costs the lives of thousands of people (Chapter 13).

8 Antarctica: the icy continent

Antarctica in perspective

Lying off the bottom of many world maps is the vast, ice-covered continent of Antarctica, bearing the largest of the two remaining ice sheets on Earth. Antarctica is a place of superlatives: it is the coldest, driest and windiest of all continents, and has the highest average elevation. It covers nearly 14 million square kilometres, which is equivalent to the area of the USA plus Mexico, or twice that of Australia, or 58 times that of the British Isles. Antarctica has a profound influence on ocean currents, climate and sea level worldwide.

Glacier ice covers all but 2.4 per cent of the continent; rock is exposed only in the Transantarctic Mountains, in coastal 'oases' and in the Antarctic Peninsula. Glacier ice reaches the coast almost everywhere. In fact only five per cent of the coastline is ice-free. The land itself can support very little life, and only sparse lichen, algae and mosses have gained a foothold on the continent. In contrast the surrounding seas are prolific, allowing numerous colonies of penguins, skuas and snow petrels to flourish during the brief summer. As for humans, probably fewer than 100 000 have visited the continent, and only a handful of hardy scientists and support staff over-winter in Antarctica.

The importance of Antarctica to the rest of the world stems from a number of factors. Firstly, the Antarctic Ice Sheet contains a phenomenal 30 million cubic kilometres of ice, sufficient to raise sea level globally by 56 metres according to one of the latest estimates. This ice represents 85 per cent of all fresh surface water on our planet. Secondly, floating ice shelves and sea ice produce cold, dense currents that flow northwards and help drive the world's oceanic circulation. Thirdly, the ice sheet is a huge heat sink and influences the climate over much of the southern hemisphere.

Mill Glacier is one of the major tributaries of the Beardmore Glacier and illustrates the wind-polished ice surface that makes travelling on foot or ski treacherous.

A helicopter of the United States Antarctic Program flies over Shackleton Glacier, central Transantarctic Mountains in support of a field party.

Antarctic glacier types

Scientists study the Antarctic Ice Sheet using surface observations, radio-echo sounding, ice-coring and satellite imagery. The Ice Sheet is not a uniform dome of ice, but comprises three unequal parts:

- the East Antarctic Ice Sheet, which contains 86.5 per cent of all the ice in Antarctica, reaching an elevation of 4000 metres in Dome Argus, 1000 kilometres from the South Pole;
- the West Antarctic Ice Sheet contains 11.5 per cent of the continent's ice (with the Antarctic Peninsula), reaches an average elevation of 2300 metres, and is punctured by a number of mountain ranges, including the highest in Antarctica, the Vinson Massif (5140 metres) in the Ellsworth Mountains;
- the Antarctic Peninsula Ice Sheet, which has several coalescing ice caps, extensive mountain terrain and highland icefields, and ice-covered offshore islands.

Figure 8.1 The component parts of Antarctica, showing the East and West Antarctic Ice Sheets, divided by the Transantarctic Mountains, the Antarctic Peninsula Ice Sheet and the principal ice shelves.

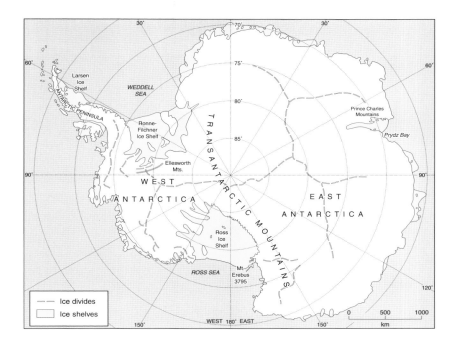

Satellite-derived image to illustrate the Antarctic Ice Sheet and its velocity distribution. The velocity is indicated by the spectrum of colours from black (zero velocity) to white (250 metres per year or more). The grey areas are ice shelves and are excluded from the analysis. It is evident that velocity is not uniformly distributed, but is focused at a number of ice streams, denoted by the white/blue areas. The black areas define ice divides where the ice is slow moving (Image derived by Jonathan Bamber.)

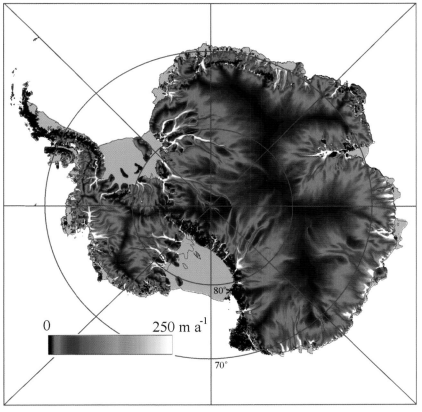

Summary of main types and lengths of coastline around Antarctica

Glacier type	Total length (km)	Percent of coastline
Ice shelf	14 110	44
Ice stream/outlet glacier	3 914	13
Grounded ice walls	12 156	38
Rock	1 656	5
Total	31 876	100

Data from Drewry, D. J. 1983. Glaciological and Geophysical Folio. Cambridge: Scott Polar Research Institute.

Within each of these subsidiary ice sheets are a number of discrete drainage basins. Arguably the largest to reach the sea directly is the Lambert Glacier system that drains into Prydz Bay and covers about one million square kilometres. Ice-flow through these basins is focused in **ice streams**, zones of fast-flowing ice, sliding on their beds, moving at speeds of several hundred metres a year. Ice streams are separated from less active ice by **shear zones**, indicated by extensive crevasses.

Where fast-moving ice streams or outlet glaciers reach the coast they either coalesce into **ice shelves**, extend beyond the coast as **glacier tongues**, or form ice-filled embayments. Ice shelves are typically 100–500 metres thick at the coast, whereas the thickness on the landward side of the larger ice shelves may be much greater. The largest ice shelves border the West Antarctic Ice Sheet, the Ronne-Filchner and Ross ice shelves each covering half a million square kilometres, but they account only for two per cent of the ice volume in Antarctica. Glacier tongues may project from coastlines for tens of kilometres, the longest occurring in the western Ross Sea. Both ice shelves and glacier tongues are the main producers of tabular icebergs that characterize the seas surrounding Antarctica.

Slow-moving parts of the Antarctic Ice Sheet reach the coast as vertical cliffs resting on the sea bottom. They are known simply as **ice walls**. They are not a major producer of icebergs, as only small

Amundsen Glacier, originating in the East Antarctic Ice Sheet, is one of many huge valley glaciers that carve their way through the central Transantarctic Mountains starting at over 3000 metres altitude and flowing into the Ross Ice Shelf at sea level.

Shackleton Glacier is one of many glaciers that originate in the East Antarctic Ice Sheet and slice through the Transantarctic Mountains, before merging in the Ross Ice Shelf. In this view we are looking across a field of heavy crevasses, down-glacier towards Mount Wade (4084 metres).

Amery Ice Shelf is a major floating slab of glacier ice, fed by arguably the world's biggest glacier, the Lambert. Here we are looking across the inner reaches of the ice shelf from Fisher Massif, a 1700-metre-high nunatak along the Lambert Glacier's western flank.

pieces of ice spall away from the cliff, gathering as rubble on the sea ice, or floating away in open water.

Ice shelves, ice streams, outlet glaciers and grounded ice walls dominate the coastline of Antarctica. Although dwarfed by the above ice masses, Antarctica also offers a wide range of other glacier types, although most of them are cold and frozen to their beds. **Ice caps**, fed by relatively moist weather systems, are common in the Antarctic Peninsula, where good examples are found on various offshore islands. **Highland icefields** draped around higher mountain massifs are a feature of the Antarctic Peninsula and North Victoria Land. **Cirque glaciers** are common on relatively low hills that cannot support an ice cap, or in relatively arid areas favoured by wind-drift snow accumulation. These glaciers may also flow to lower levels as **valley glaciers**. Spectacular and unusual valley glaciers are a feature of the Dry Valleys of Victoria Land. Most have high frontal ice cliffs, preventing ready access onto them. Glaciologists have excavated several tunnels in the base of such cliffs, and have found that, despite sub-zero temperatures (typically −15 °C or lower), these glaciers are in fact eroding the underlying bedrock and frozen sediment, and then transferring the material to the margin, so creating moraines. This observation is contrary to the widely held view that glaciers frozen to their beds protect the landscape from erosion.

How thick is the Antarctic Ice Sheet?

The technique of airborne radio-echo sounding, where radio waves are used to discriminate between the ice surface, internal structure and the bedrock, has revolutionized our understanding of the ice sheet. The technique was discovered almost by accident when radar was being used to monitor the height of a low-flying aircraft above the ground. The observers found that the radar waves passed through the ice, recording bedrock, rather than the ice surface. Extensive radio-echo sounding surveys were first undertaken in the 1970s by British and American scientists using transmitters and receivers mounted beneath the wings of a Hercules aircraft. Today, smaller aircraft, notably Twin Otters, are used and the surveys, using the latest Global Positioning System technology, are accurate to within a few metres.

Radio-echo sounding profiles have been undertaken over much of the ice sheet. We can calculate from these results that, if the ice is spread out evenly over the continent, it would be 2160 metres thick. In fact, the deepest ice is 4776 metres in the Astrolabe Subglacial Basin. Deep ice also occurs in the Bentley Subglacial Trough where the bedrock is 2555 metres below sea level. The ice sheet buries entire mountain ranges, such as the 3000-metre-high Gamburtsev Subglacial Mountains in East Antarctica. In other areas, just the tips of mountains project through the ice as **nunataks**, increasingly so towards the coast. A major feature of the Antarctic continent is a huge range known as the Transantarctic Mountains; although much of this range is buried, it does 'hold back' much of the East Antarctic Ice Sheet and prevents it from draining into the major ice shelves. Spectacular trunk glaciers break through the range to help sustain the ice shelves, such as the Beardmore and Axel Heiberg glaciers that provided the first overland routes to the South Pole. The mountains that project through the ice sheet in this range include several more than 4000 metres high, the highest being Mount Kirkpatrick (4528 metres). Impressive as the Transantarctic Mountains are when viewed from the coast of Victoria Land, on the ice sheet side they appear as mere dimples.

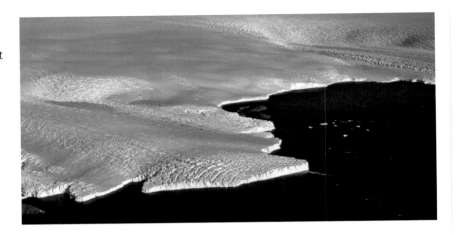

Most of the discharge from the East Antarctic Ice Sheet is via ice streams, zones of fast-flowing ice between ice that is frozen to the bed. Crevasses mark the boundary with slow-moving ice, and the ice streams themselves are heavily crevassed and show how the ice is drawn-down by faster flow. These two examples occur on the coast of Wilkes Land.

Ice sheet dynamics

Glaciers everywhere are sustained by snowfall, and the balance between profit and loss of ice determines how fast they flow and whether they recede or advance. Usually glaciers flow from a high altitude zone of accumulation to a low altitude zone of ablation dominated by melting. The situation in Antarctica is somewhat different. In the interior of the ice sheet accumulation is slight – only a few centimetres a year. Some medium-to-low elevation areas in the interior experience net loss of mass, as a result of the combined effects of wind and sublimation (direct evaporation without melting). However, towards the coast, there is commonly an increase in snowfall, leading to net accumulation close to sea level. Consequently, many ice shelves are composed substantially of ice derived from coastal snowfall. In addition, some ice shelves gain mass by the freezing of marine water to their undersides. Therefore, a substantial proportion of the ice reaching the coast of Antarctica is not derived from the interior. Furthermore, loss of ice from Antarctica is not by the usual process of melting, but by calving of icebergs into the sea. The relative proportions of ice delivery into the ocean are as follows: ice shelves 62 per cent, ice cliffs 16 per cent, and ice streams/outlet glaciers 22 per cent.

In view of these complexities in accumulation and ablation, it is not surprising that the flow dynamics of the ice sheet are difficult to

Constant freeze-thaw of ice in spring causes huge icicles to form, as here at the front of Mackay Glacier, western Ross Sea. The dark-coloured objects on the sea ice are Weddell seals.

Floating extensions into the sea of glaciers are termed glacier tongues. Here, the Erebus Glacier tongue, with its saw-toothed flanks, flows off Mount Erebus into McMurdo Sound. Smooth snow-covered sea ice surrounds the tongue on all sides.

Ice shelves are floating slabs of glacier ice, and commonly show advance and recession unrelated to climatic change. They undergo slow build-up and expansion over many decades, followed by calving of huge icebergs within a few months. This ice shelf in Princess Elizabeth Land, although small, shows in miniature how some of the exceptionally large icebergs form.

Opposite. Two views of the Ekstroem Ice Shelf, East Antarctica. The upper surface near its edge is marked by a knobbly icy surface – the result of storm waves throwing up spray that freezes as a crust on the surface. The lower photograph illustrates the shelf edge, marked by an undercut cliff, rising about 30 metres above water level, while a few more hundred metres of cliff extends below the water-line.

unravel. Nevertheless, scientists have identified some key components of glacier flow. They believe that many ice streams flow rapidly because the bed on which they rest is of soft sediment. The bed behaves like a slurry or lubricant, allowing fast flow. The ice shelves themselves also play an important role. If they are stable, they effectively buttress the ice sheet, preventing it from discharging even more rapidly into the sea. The concern is that if ice shelves disintegrate, there will be nothing to stop catastrophic discharge of ice from the interior into the ocean, raising sea levels globally. In this respect, the West Antarctic Ice Sheet is the most vulnerable, as much of it is grounded below sea level, so if the buttressing ice shelves were removed, the interior ice would become buoyant and collapse into the sea. Much further research is necessary to ascertain whether this doomsday scenario is realistic.

A buried landscape

Radio-echo soundings have yielded vital information concerning the bedrock elevation beneath the ice sheet. The land beneath East Antarctica is mostly above present-day sea level, but mainly below 1000 metres altitude. Exceptions are several subglacial mountain ranges, spectacular 'alpine-style' nunataks as in Dronning Maud Land, and isolated rolling uplands bounded by glacially carved cliffs as in the Prince Charles Mountains. There are also several subglacial basins below sea level, such as the Wilkes-Pensacola Basin that stretches across Antarctica on the inside of the Transantarctic Mountains.

By contrast, the bedrock in West Antarctica is mainly several hundred metres below sea level, but is very irregular, being punctuated by mountain ranges, notably the Ellsworth Mountains. The Antarctica Peninsula has bedrock largely above sea level, and has numerous nunataks over 1500 metres high projecting through it.

All this ice on Antarctica has effectively depressed the crust. If it were to melt, over thousands of years the crust would rebound. In East Antarctica a rise of the crust of 1000 metres is possible, so

exposing most of the subglacial basins above sea level. In West Antarctica rebound of around 500 metres is anticipated, but this would leave the region still mostly below sea level.

The iceberg factory

Calving of icebergs is the principal means by which the Antarctic Ice Sheet loses mass. The scale of this process far outweighs any other form of ablation of glaciers worldwide. Iceberg production influences the character of the Southern Ocean, and it is believed by some scientists that extreme events can alter the Southern Hemisphere climate.

Antarctic icebergs are characteristically tabular – thick flat-topped slabs of ice up to 500 metres thick and often many kilometres long. The exceptional depths reached by the inner continental shelf means that icebergs are able to float away freely. Since nine-tenths of their thickness lies below water, their drift is influenced more by oceanic currents than by wind action, in contrast to that of sea ice. It is thus common to observe icebergs ploughing through sea ice in a totally different direction to that of the wind. Decaying icebergs behave in a fickle manner, as their centre of gravity shifts on melting. They may alter course without warning, so mariners have to be constantly on the alert.

Both ice shelves and glacier tongues produce tabular icebergs. Smaller bergs of irregular shape calve from grounded ice cliffs and from the decay of tabular bergs. Many icebergs become grounded on submarine banks, and decay within Antarctic waters. Others float freely and sail far beyond the Antarctic continental shelf. There is a general movement of icebergs west around the East Antarctic coast. However, the Antarctic Peninsula blocks passage of icebergs in the Weddell Sea region forcing them northwards in the so-called Weddell Sea Gyre until they are released from Antarctic waters. Sightings of icebergs as far north as Cape Horn (South America) and Cape of Good Hope (South Africa) were not uncommon in the nineteenth century.

As tabular icebergs decay, they begin to tilt, leaving a succession of former waterlines on their flanks. Many become completely

Satellite image of the Larsen Ice Shelf. The shelf is shown undergoing catastrophic collapse during February and March 2002, following decades of rising temperatures in the Antarctic Peninsula region. Multi-angle imaging radio spectrometer instrument aboard TERRA satellite. (Image courtesy of NASA, from http://visibleearth.nasa.gov/cgi-bin/viewrecord?12416.)

upended, revealing the full thickness in the horizontal dimension, as indicated by the annual stratification that is commonly observed in the seaward cliffs of ice shelves. Some icebergs 'turn turtle' and reveal the underside of the source glaciers. Small icebergs of high density called **growlers** may be almost invisible and represent a major hazard to shipping. When icebergs break up the product is known as **bergy bits**.

Calving of large icebergs is a subject that seems to fascinate the media, the thinking being, erroneously, that such events herald the demise of the ice sheet. Ice shelves produce the largest icebergs, and some detached from Antarctic ice shelves are many kilometres across. In 1987 satellite imagery recorded the breaking away of an

Where the East Antarctic Ice Sheet fails to reach the coast, perhaps because it is blocked by mountains, conditions are so dry that there is less precipitation than in tropical deserts. Dry polar desert conditions are found in the Dry Valleys of Victoria Land, and sand dominates the once-glaciated terrain. Here, in Victoria Valley, a small lobe of ice, known as the Victoria Lower Glacier, flows inland from a coastal icefield, terminating in this dusty basin. The limited meltwater also drains inland before evaporating.

iceberg from the 800-kilometre-wide Ross Ice Shelf that was nearly 160 kilometres long, and occupied an area of over 6250 square kilometres. It was said by a spokesperson of the US National Science Foundation to contain enough water to supply Los Angeles for 675 years! The largest iceberg ever recorded, named B15, broke away from the Ross Ice Shelf in March 2000; it measured 295 kilometres in length, had an estimated thickness of 100–350 metres, and covered an area of 11 000 square kilometres, equivalent to an area the size of Jamaica or half the size of the American state of Massachusetts. The iceberg then exerted pressure on another part of the ice shelf causing another berg, 110 kilometres long, to break off. B15 broke into two after seven weeks.[1]

Other recent calving events include the calving of several icebergs

[1] Antarctic Meteorological Research Centre, University of Wisconsin, Madison.

Most glaciers in the Dry Valleys are cold-based, but nevertheless rework the underlying sediment and rock, while trickles of meltwater cause a certain amount of fluvial erosion. Hughes Glacier is a self-contained mountain glacier high above Lake Bonney, the surface of which is frozen and saline; like many lakes fed by glacial meltwater in this region, the lake has no outlet.

covering a total area of 13 000 square kilometres, from the Filchner Ice Shelf in 1986, the break-out from the Larsen Ice Shelf which produced a single iceberg covering 9000 square kilometres in 1996, and the calving of the second longest iceberg (250 kilometres) from the Ronne Ice Shelf in May 2000. Large icebergs have only been recognized and tracked since the advent of modern satellite imagery, so B15 is unlikely to have been the biggest ever. Even as long ago as 1927, an iceberg 180 kilometres long was observed as far north as the Scotia Sea; it was probably several years old, having travelled between 500 and 2000 kilometres.[2]

All icebergs should be given a wide berth by shipping. Not only can they turn over or break up without warning, but many have under-

[2] Antarctic Meteorological Research Centre, University of Wisconsin, Madison.

water projections called keels, that can rip open the hull of a ship as would collision with a reef.

Antarctic icebergs are purveyors of considerable amounts of rock debris into the Southern Ocean and beyond. Whereas most Southern Ocean floor sediment is muddy or rich in microscope organisms such as diatoms, isolated boulders, pebbles or clumps of coarse-grained debris also occur. If the rock types can be identified, then it may be possible to locate the source area. The number of Antarctic icebergs revealing debris is, however, limited. The reason for this is that in tabular bergs most debris will be in the former basal zone of the source glacier. There are some exceptions, notably where icebergs are derived from northern areas of the Antarctic Peninsula. Glaciers in this region carry significant amounts of supraglacial (surface) debris, and even remnants of medial moraines may be transported on top of a berg before it melts.

Among the earliest glaciological studies in Antarctic were those undertaken by the British expeditions of Scott in the early years of the twentieth century. Scott's second but ill-fated expedition established this base at Cape Evans in McMurdo Sound, seen during the tail-end of a blizzard. Barnes Glacier in the background is a cliff grounded on the sea floor.

Icebergs are often used by animals as resting places. Agile Adélie penguins often shoot out of the water and use their flippers to get onto reasonably high icebergs. Seals sometimes haul out onto thin flat icebergs, although they tend to prefer sea ice. Flying birds, such as gulls, terns and petrels, also may temporarily perch on an iceberg.

Subglacial lakes

Surprisingly, there are many lakes beneath the Antarctic Ice Sheet, none of which has been observed directly by humans. Although there had been hints that subglacial lakes were present, such as the presence of extensive flat areas at the surface of an otherwise gently sloping ice sheet, the main breakthrough in their recognition came when radar soundings were undertaken in the early 1970s. A detailed examination of the radar data by American, British and Russian scientists revealed a number of small subglacial lakes, as well as the internal layering of the ice sheet, culminating in the discovery of Lake Vostok in 1977 beneath the Russian station of the same name in the heart of the East Antarctic Ice Sheet. As of now, over 70 subglacial lakes have been identified beneath the Antarctic Ice Sheet. By combining accurate mapping, using an altimeter on the European Space

Beardmore Glacier, slicing its way through the Transantarctic Mountains, is one of the major overland highways to the South Pole from the Ross Ice Shelf. The route was pioneered by Shackleton in 1908, and followed by Scott in 1911–12. Mount Kyffin (1604 metres) marks the entrance to the glacier, which here is heavily crevassed as it joins the ice shelf, illustrating the severe challenges faced by the early explorers.

Small manoeuvrable vessels such as the *James Caird* are often used to enhance hydrographical and marine surveys undertaken by their larger host vessels in iceberg-infested waters, such as *HMS Endurance*. Snow Hill Island, northern Antarctic Peninsula is in the background.

Tabular iceberg in the early stages of decay off Mawson Station, Mac. Robertson Land, East Antarctica. Waves have excavated two caves at the waterline in one of the bergs. The sea ice around the bergs is freshly formed and heralds the onset of winter conditions in March.

Agency's ERS-1 satellite, and the radar data, the dimensions of Lake Vostok have been determined. It has proved to be surprisingly large: 230 by 50 kilometres, covering an area of about 14 000 square kilometres. The depth of this lake is approximately 500 metres beneath Vostok Station, yet it lies below 3700 metres of ice.

Vostok Station at the surface of the East Antarctic Ice Sheet has another even more important claim to fame, in that an ice core obtained here has provided us with the world's longest continuous high-resolution climatic record (as discussed in Chapter 15). In the context of the subglacial lake story, the core has provided us with important information about the age and life forms living therein. The oldest true glacier ice recovered at Vostok is around half a million years old, which means that the water in the lake may be at least this old. Indeed, since layers of lake ice have frozen onto the base of the ice sheet here, the lake itself could be much older, and some scientists have speculated that it could be several million years old. The possibility of finding life forms in this lake that have evolved in complete isolation from the rest of the planet has excited both scientists and the public. Admittedly, conditions are not exactly favourable to life because of the high pressures (35.46 MPa or 350 atmospheres), low temperatures ($-3\,^{\circ}$C) and permanent darkness. However, there is likely to be oxygen, released from the glacier ice, while chemical sources may drive biological processes. Neither Lake Vostok nor the other subglacial lakes have yet been plumbed, owing to the concern about possible contamination (as well as cost). However, the lake ice layers near the base of the Vostok ice core have been examined for life forms, and low concentrations of bacteria have been found, demonstrating the potential of finding new life.

Much of the interest in obtaining information about life in subglacial lakes is driven by the American space organization, NASA, in connection with its plan of a mission to Europa, one of the moons of Jupiter. Europa has an ice crust several kilometres thick, with a liquid ocean beneath, a potential location for extra-terrestrial life. A trial scientific programme on a lake such as Vostok is therefore deemed appropriate. Apart from undertaking a full site survey, the

The East Antarctic Ice Sheet produces vast quantities of tabular icebergs, some the size of small countries. In this aerial view off the coast of Coats Land, eastern Weddell Sea, we see a number of recently calved icebergs, each measuring a few hundred metres in length. Numerous intersecting crevasses on their upper surfaces are evident, as are smooth areas of sea ice and ice debris resulting from collision of adjacent bergs. Note also how the bergs tilt; this indicates the early stages of disintegration.

main challenge to explore Lake Vostok is to develop a means of sampling without contaminating the water. The plan is to begin with hot-water drilling to about 200 metres above the lake surface, and then lower a probe with scientific instruments tethered to the ice surface. At this depth, an instrument called a 'cryobot' would be released, and melt its way down through the ice, sealing the hole above itself as it descends. It would have to be decontaminated at an early stage of this process. On reaching the lake surface it would act as a remote station for gathering data. Another instrument, a 'hydrobot' would be released by the cryobot and its movement controlled from the ice surface nearly four kilometres above. The hydrobot would then explore the lake, taking samples of water and sediment. Samples would need to be maintained at the high pressure levels experienced in the lake as the cryobot returned them to the ice surface. Such developments may seem to be the stuff of science fiction, but already by November 2001 the first experiments with a prototype cryobot had been undertaken on a small glacier in Svalbard.

In reality, the remoteness and size of Lake Vostok makes it far from ideal as a first target. Given that there are many other subglacial lakes, likely to have similar potential for life, plans have been mooted to explore smaller lakes close to existing research stations with good infrastructure, such as the newly rebuilt South Pole Station, served by frequent flights from McMurdo on the shores of the Ross Sea.

Is the Antarctic Ice Sheet growing or shrinking?

We have, as yet, no simple answer to this question, as mixed messages are emerging from investigations in the Antarctic. In the Antarctic Peninsula, temperature records show a 0.5 °C increase every decade since 1947. At the same time several large ice shelves have disintegrated, including the Wordie and Larsen. James Ross Island in the north recently separated from the mainland by the disintegration of the Prince Gustav Ice Shelf. Furthermore, many of the land-based glaciers have thinned and receded substantially, especially in the northern Antarctic Peninsula. Some people have argued that these events are precursors to the disintegration of the entire ice sheet.

In the case of ice shelves, it is known that they have disappeared in the past. For example, sediments beneath the former Prince Gustav Ice Shelf indicate that ice-shelf collapse took place several thousand years ago. Scientists have also shown that ice shelves grow slowly through time, and then are subject to short-term calving events. For example, the western front of the Ross Ice Shelf is at its most advanced position in recorded history.

Elsewhere, many local glaciers, as in the Victoria Land Dry Valleys, are slowly advancing in response to increased snow precipitation. Herein lies a clue to the future behaviour of the ice sheet. Although peripheral areas will probably continue to recede, the more southerly regions may grow. Warming in high Antarctic latitudes will be insufficient to raise temperatures above freezing initially, but will bring in its wake increased precipitation as snow. Computer models have shown that Antarctica can sustain a 5 °C temperature rise before going into decline. Global climatic trends do, however,

Drifting far beyond the Antarctic coastline in the Southern Ocean are bergs in various stages of decay, some maintaining their tabular form, others overturning and acquiring irregular shapes.

suggest that temperature rises of this order will occur over the next century.

Predicting the future

At present we cannot be sure of Antarctica's contribution to sea-level rise, in contrast to the other smaller ice masses. We do not know whether, overall, the ice sheet is growing or shrinking. Scientists need to gather mass-balance data over several decades before being certain, one way or the other, whether the Ice Sheet is changing. For this, satellite data are essential and need to be sufficiently refined to monitor centimetre-scale changes of the elevation of the ice sheet surface.

However, using powerful computers, numerical modelling can be used to predict the response of the ice sheet to climatic change. Modellers use a wide range of glaciological data to 'drive' their models, such as present-day climate, ice sheet dynamics, iceberg calving rates and characteristics of the bed. However, model results are only as good as the data used, and they do not provide definitive solutions. Nevertheless, models can be tested against the geological record. If model results are made to match geological evidence, then plausible ice sheet scenarios are achievable for the future. This approach has been successfully applied to former ice sheets in the Northern Hemisphere. It is now being applied to predicting the future of the Antarctic Ice Sheet.

Within our grasp, then, is the means to predict the scale of changes to the ice sheet in response to increases in temperatures resulting from greenhouse gases. We should, for example, be able to say what sized ice sheet is stable for a given climatic regime. The big challenge is to be able to determine how quickly the ice sheet will respond to temperature rises that are unprecedented in their rapidity in geological time.

Figure 8.2 The Antarctic Peninsula, illustrating the collapse of ice shelves in the late twentieth and early twenty-first centuries. (Adapted from an article by J. Kaiser in *Science*, **297**, p.1495).

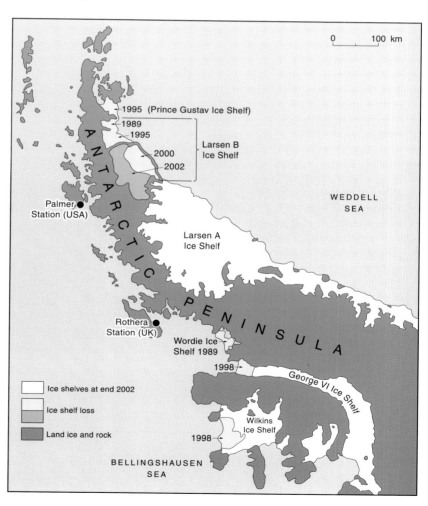

9 Glaciers and volcanoes

Volcanoes are mostly situated at the boundaries of tectonic plates, such as destructive plate margins where oceanic crust is forced beneath continental crust, or at constructive plate margins where new oceanic crust is being created. The Pacific Ocean is bordered by destructive plate margins, and so many volcanoes occur around its rim. This so-called 'Ring of Fire' runs through the Andes, across the Pacific to New Zealand, through Japan and the Kamchatka Peninsula of Russia, back across the Pacific to the Aleutian Islands and mainland Alaska, and down through the Western Cordillera to Mexico and the Andes. In contrast, the Atlantic Ocean only has volcanoes in a series of islands along the Mid-Atlantic Ridge.

The association of glaciers with these volcanoes is linked to either high altitudes, as in the tropical Andes, or to high latitudes as in Iceland, Jan Mayen or Antarctica. Glaciers on high mountains are generally thin, but an eruption can have devastating consequences because the combination of melting ice and loose debris can generate unpredictable fast-flowing mudflows called **lahars**. Where the ice is several hundred metres thick, as in Iceland, large volumes of meltwater are generated during subglacial eruptions. In Antarctica, geologists have documented subglacial eruptions in the rock record, but these eruptions have had little obvious impact on human civilization. Subglacial eruptions on Mars have even been postulated, where scientists have inferred that ice sheets once were much more extensive than the small ice caps of today.

The products of glacier-covered volcanoes

Despite the occasionally spectacular eruptions that occur beneath glaciers, together with the accompanying flooding, we know relatively little about the processes involved, and how the resulting

Mount Rainier (4392 metres), Washington State, USA carries the largest glacierized area and ice volume within the contiguous United States of America. This substantial ice reservoir is a serious potential threat to cities to the west of the volcano should volcanic activity resume in the future. Avalanche lilies are in the foreground in this summer view taken from the south.

The near conical outline of Osorno in Chile. The last known eruption of Osorno took place in 1869. Lahars contain volcanic material from mud to boulder size and are mobilized by snow and ice melt during an eruption or, as in this case, by intensive summer ablation.

glacial sediments and volcanic rocks are produced. This is particularly true of eruptions beneath thick glaciers. The products of high glacier-capped Andean volcanoes are similar to those of explosive volcanoes without ice. Typical deposits of volcanoes at destructive margins are ash falls, sticky (**andesitic**) lavas, lahars and welded rock fragments that travel long distances from the eruption site. If this type of volcano is glacier-capped, then the main difference is the greater abundance of meltwater-generated lahar deposits.

The main product of subglacial volcanoes in areas of constructive plate margins is relatively 'runny' black **basalt**. Beneath thick ice, meltwater is often ponded, so basalt lava deltas are produced, and the basalt extruded into water occasionally develops pillow-shaped structures a metre or so in diameter. Disaggregation of the lava into angular fragments of many sizes produces beds of breccia tens of metres thick; the resulting rock is called **hyaloclastite**. The volcanic rocks may be associated with a glacial deposit called **till** that has melted out from the base of the glacier. There may also be striated pavements, representing glacial erosional phases between eruptions. Compared with the conical form of a typical volcano, subglacial volcanoes have steep sides and a flat top; many of these can be seen in formerly ice-covered areas of Iceland, where they are known as **stapi**. Similar volcanic forms occur in British Columbia under the name **tuyas**.

Although the best known subglacial volcanoes are in Iceland, many also occur in Antarctica, mainly in the Antarctic Peninsula region and beneath the West Antarctic Ice Sheet, together with a few along the western margin of the Ross Sea. Those beneath the ice sheet may pose a hidden threat to civilization on a global scale, since a major subglacial eruption could destabilize the ice sheet, and induce rapid ice discharge into the Southern Ocean. Total collapse could result in a sea-level rise globally of six metres.

Volcanoes which erupt subglacially have a distinct form, with a flat or slightly domed top. Herðubreið in northern Iceland erupted beneath the ice sheet which once covered most of the island during the last glaciation. The thickness of the ice sheet is indicated by the height of the steep cliffs. The meltwater torrent in the foreground is an outflow from Vatnajökull, the largest of the remaining icecaps in Iceland.

Glaciers associated with the 'Ring of Fire'

In 1986 the Colombian volcano Nevado del Ruiz erupted. On a geological scale, the event was small. Only small amounts of ash were deposited, mainly on the eastern side of the volcano, and little physical damage was caused. Nevertheless, around 30 000 Columbians lost their lives because the snow- and ice-capped Nevado del Ruiz had triggered the most tragic of all twentieth century volcanic catastrophes. When the eruption started, glacier ice melted and the water carried away enormous quantities of old and new ash, creating a lahar. This lahar tore down the mountainside into the densely populated valley below, completely burying the town of Armero in minutes.

In the Andean countries Columbia, Ecuador, Peru, Bolivia and Chile, volcanoes and glaciers combine to form some of the most spectacular, but also potentially dangerous, landforms. The extent of glacierization illustrates the effect of temperature and precipitation upon the altitude of the snowline and therefore the presence of glaciers. In the wet tropical zones of Columbia and Ecuador volcanoes only 5500 metres high carry glaciers. Throughout Peru and on into northern Chile the altitude of the snowline actually increases despite

Past evidence of subglacial eruptions on James Ross Island, near the northern tip of the Antarctic Peninsula, is recorded in these volcanic rocks near Whisky Bay. Stretching back over some 10 million years, these volcanic rocks (called hyaloclastite) were produced as lava made contact with meltwater and broke up explosively into fragments.

decreasing temperatures, because of the lack of snowfall. Around the Tropic of Capricorn, east of Antofagasta in Chile, volcanoes stand much higher than 6000 metres but are entirely free of glaciers. South of the Chilean capital Santiago, the snowline rapidly falls, as temperatures decrease and snowfall increases. South America's southernmost volcanoes are found east of the towns of Valdivia and Puerto Montt. Here, glaciers crown all volcanoes between 2500 and 3000 metres high.

Arguably the most beautiful of these is Volcan Osorno. Its brilliant white glaciers and remarkably symmetrical cone reflect in the dark green waters of Lago Todos los Santos, one of the many glacial lakes in southern Chile. One of its neighbours, Volcan Villarrica, is a major tourist attraction. In a nearly permanent state of moderate eruptive activity, this volcano attracts thousands of people. Taking part in one

Volcanoes that have been inactive for a long time are subject to erosion, such as by glaciers, as here on Tronador on the border between Argentina and Chile. This volcano has long since lost the cone shape that is typical of composite central volcanoes, as cirque and valley glaciers have eroded substantial parts of the volcanic edifice.

of the guided tours to its summit, visitors are able to peer inside the crater and sometimes glimpse small-scale eruptions of magma. Trekking to the summit necessitates crossing a small but steep crevassed glacier. Since many people have become lost on the mountain, the local authorities now prevent unqualified trekkers from attempting independent climbs to the summit crater.

Villarrica has not always been the tame volcano it is today. As recently as December 1984, lava flows from the summit crater cut deep canyons into the glaciers and triggered lahars. In prehistoric times hot pyroclastic avalanches covered as much as 3000 square kilometres of terrain. During such powerful eruptions, probably all glacier ice was lost.

The glacierized volcanoes of South America are perhaps the most dramatic on Earth, but there are other impressive examples around the Pacific Ocean. New Zealand's North Island has several impressive volcanoes, the highest of which, Ruapehu (2797 metres), provides the island's premier skiing area. Ruapehu receives heavy winter snowfall, and carries small glaciers within and around its crater. On Christmas Eve in 1953 an outburst from the crater lake, supplemented by the melting of snow and ice, generated a lahar which swept away a railway bridge shortly before the Wellington-Auckland express was due to pass by. The locomotive and five carriages plunged into the torrent and 151 people lost their lives. Ruapehu is monitored very carefully these days, as its crater lake is a permanent hazard. The volcano's last eruption began on 24 September 1995. At this time the spring skiing season was still in full swing in the two resorts on the volcano's western flank. From the ski slopes people watched in awe as Ruapehu ejected billowing clouds of ash and catapulted huge blocks of glacier ice and rocks hundreds of metres into the air. Lahars once again raced down the mountainside but fortunately this time there were no casualties. Ash falls eventually rendered the ski pistes useless, and the local tourist industry finally declared the event an economic catastrophe.

Travelling north around the Ring of Fire in the western Pacific we reach volcanoes with glaciers in Russia's Kamchatka Peninsula,

Glaciers around Ruapehu (2797 metres), North Island, New Zealand. Failure of an ice dam resulted in a massive debris flow (lahar) which swept away a bridge and led to the Christmas 1953 railway disaster. During eruptions of Ruapehu (as in 1995) the lake is also the cause of many lahars.

which attain altitudes of more than 4500 metres. Many more ice-clad volcanoes are found further east, on the Aleutian Islands and on the southern mainland of Alaska. Most of these are far-removed from centres of population.

In the contiguous United States, Washington State's Cascade Range is crowned by a string of glacierized volcanoes. Perhaps nowhere have the effects of a volcanic eruption been studied more intensively than on Mount St. Helens in the Cascade Mountains of Washington State. The huge eruption on 17 May 1980 reduced the mountain's height from 2949 to 2549 metres and most of its glaciers disappeared. The eruption began on 27 March 1980, when, after several days of earthquake activity, small ash and steam explosions penetrated the summit ice cap, ending the volcano's 123-year dormancy. During April and early May magma appeared on the

The eruption of Mount St. Helens, Washington Sate, USA in 1980 resulted in the destruction of its near-perfect conical form and the glaciers on its flanks. It also reduced its height from 2950 metres to 2549 metres. Twenty years later a new glacier is forming within the crater around the central dome. The glacier is grey because it is covered by ash from the crater walls. Note Mount Rainier, the extensively ice-covered volcano in the background.

previously symmetrical cone, producing a noticeable bulge on its north flank. However, only a few ice avalanches were generated at the time, despite the over-steepening of the mountainside, and the glaciers remained more or less intact.

The main eruption on 17 May resulted from slope failure on the bulging north side of the mountain, which created a huge landslide. Sequential photographs of the event show an enormous ice avalanche tearing down the mountain, probably representing most of the accumulation area of Forsyth Glacier, which had been thrown off the summit area as the explosion commenced. During the eruption 70 per cent of the mountain's glacier volume disappeared, and today only small glacier remnants are left on Mount St. Helens' outer slopes. However, an unusual, small glacier has formed inside the crater. It is mainly fed by snow avalanches from the steep walls of the crater's interior. As the terrain is dusty, the glacier is very 'dirty' and barely noticeable.

Whereas only a few square kilometres of ice covered Mount St. Helens before the 1980 eruption, its majestic northern neighbour, Mount Rainier (4391 metres), has more than 90 square kilometres of glaciers, including Emmons Glacier, the largest in the United States outside Alaska. Because of the erosive power of its large glaciers,

Mount Rainier's cone is less perfect than that of Mount St. Helens before its last eruption. Although Mount Rainier has erupted repeatedly in the last thousand years, it has been quiet since the mid-1800s. This is just as well, as the distribution of lahars formed in the last 10 000 years gives an indication of the devastation that may occur in the event of a major eruption in the future. The lahars range in length from a few kilometres to 110 kilometres (extending as far as the outskirts of Tacoma), and some fill valleys to a depth of many tens of metres.

On another volcano, Mount Shasta in California, are the southernmost glaciers in the United States. In Mexico the volcanoes Popocatépetl and Citlaltépetl are crowned by glaciers because of their considerable elevations, 5452 and 5700 metres respectively. Currently (in 2003) Popocatépetl is in an active eruptive phase that started in 1996. Its glaciers are bombarded by ash-falls and volcanic bombs, but some ice surprisingly still remains.

Glaciers on the Mid-Atlantic Ridge

The Mid-Atlantic Ridge, which stretches the length of the South and North Atlantic oceans, emerges in a few places above sea level as volcanic islands. Two of these islands, Iceland and Jan Mayen, occupy sub-Arctic and Arctic latitudes respectively, and have extensive glaciers.

Iceland has the best-documented history of subglacial volcanic activity and corresponding floods. Ten per cent of this country is covered by glacier ice, most of which lies astride this volcanically active rift zone. Here the Earth's crust is spreading at a rate of several centimetres a year. Iceland is an unusual place, as volcanic activity here is more vigorous and productive in terms of lava output than anywhere else along the Mid-Atlantic Ridge, largely because a so-called hotspot enhances the heat flow from within the Earth's mantle.

By far the largest of Iceland's ice caps is Vatnajökull, which has an area of 8100 square kilometres and is mainly 400–700 metres thick.

Near its centre is a volcano known as Grímsvötn, with its broad ice-capped caldera, which has erupted frequently in historical times, most recently in 1998. Usually it is airline pilots on routine flights who notice the ash column penetrating the ice. However, in recent years an extensive seismic network has allowed Icelandic scientists to monitor volcanoes, even those covered by ice.

It was by using these geophysical methods that the start of a larger than normal subglacial volcanic eruption was recorded on 30 September 1996 at a site subsequently named Gjálp, about five kilometres north of Grímsvötn. An earthquake of magnitude 5.4 on the Richter scale marked the initiation of the eruption. Thousands of earthquakes took place under the northwestern section of Vatnajökull. On 1 October two large depressions or cauldrons formed in the surface of the ice cap. Subglacial melting led to a slow collapse of the overlying glacier ice that initially was 550–750 metres thick. New cauldrons developed, indicating a 6-kilometre-long fissure eruption. Simultaneously, water from the eruption site filled the subglacial cauldron of Grímsvötn, forcing the 200- to 250-metre-thick ice cover to rise. On 2 October the eruption penetrated the ice cap. The ash column eventually reached nine kilometres in height and was accompanied by a spectacular lightening display. The caul-

The subglacial eruption of the newly   (a)
named volcano Gjálp, in Iceland in autumn 1996 was one of the most spectacular glaciological events of the last decade. The volcano, initially erupted beneath the ice cap, Vatnajökull (a), and created a large lake before sending a large plume of ash skywards. The subglacial lake eventually burst out beneath the glacier, emerging at the terminus of the outlet glacier Skeiðarárjökull as a huge flood (b). The aftermath of the flood shows a huge channel carved into the ice margin that dwarfs the people on the flood plain (c). (Photographs courtesy of Magnus Gudmundsson (a, b) and Andrew Russell (c)).

drons were now all linked by a shallow trough associated with sub-glacial meltwater flow. On 4 October, explosive eruptions seemed to be occurring through a water body at least 50 metres deep.

Explosive volcanic activity ceased on 13 October, but by then the eruptive fissure was marked by two large steaming ice cauldrons. Water was observed flowing rapidly via an ice canyon with 150-metre-high cliffs into a large tunnel. By this stage a thin layer of ash

(b)

(c)

Opposite. The southernmost active volcano in the world is Mount Erebus (3795 metres), and has a lava lake at its summit, giving off occasional puffs of steam. The volcano, seen here from New Zealand's Scott Base and from the air, is almost completely ice-covered.

mantled much of the ice cap. Clearly, the subglacial eruption generated enormous quantities of meltwater. Because the Grímsvötn caldera was known to produce jökulhlaups (outburst floods) on a four- to six-year cycle, Icelandic scientists monitored the depth of water in the caldera. Once the water reached a critical level, based on previous experience, a jökulhlaup was predicted. However, on this occasion the water continued to rise to the unprecedented level of 120 metres, 55 metres above normal. Finally, the jökulhlaup began on 5 November: 4.7 cubic kilometres of water bursting out onto the outwash plain near sea level after flowing under the southern outlet glacier of Skeiðarárjökull. The scale and effects of the flood, which briefly made the outlet flood the second largest river in the world in terms of discharge, are described in 'Ice and water' (Chapter 7).

Volcanoes in Antarctica

Volcanoes can be found in two main glaciological situations in Antarctica. The first group of volcanoes has just a thin cover of ice (less than 150 metres). Because of very active glacial erosion, they typically occur as remnants that retain few clues of their original form. When such volcanoes erupt, meltwater readily drains down the slopes and the products are mainly **pyroclastic deposits**, volcanic **ejecta** (ash, bombs and **tephra**) and lava. Glacial erosional surfaces and interbedded glacial deposits are associated with each eruptive phase. Deception Island in the sub-Antarctic South Shetland Islands shows the most recent signs of large-scale eruptions, with a major event having taken place in 1969.

The much higher Mount Erebus (3794 metres) in the western Ross Sea is actually formed where the crust is being stretched, and is dominated by basaltic lava and ejecta. The Ross Sea has several down-faulted rift basins, whereas its western flank is dominated by the Transantarctic Mountains that represent the 'shoulder' of the rift system. Mount Erebus is almost entirely glacier-clad, yet contains a lava lake in its summit crater, thereby providing one of the greatest natural temperature contrasts within a few metres anywhere on

Earth (around 1000 °C for the lava and −60 °C or lower for the winter air temperature). Since its discovery in 1841 by James Ross, no-one has witnessed a large-scale eruption. There have been a few short phases when ash has been ejected, but a plume of steam is much more typical.

Volcanoes that formed below thick ice in Antarctica are more numerous that cone-shaped volcanoes. Many are to be found in the Antarctic Peninsula and adjacent islands, and also in West Antarctica. However, most are buried beneath the ice and there have been no reports of actual eruptions. James Ross Island, east of the northern tip of the Antarctic Peninsula, is exceptionally well endowed with multiple phases of subglacially erupted volcanic rocks, the ice having thinned sufficiently today to reveal the form and deposits of the volcanoes.

There are many volcanoes in the Antarctic Peninsular region, but most are completely ice-covered and are thought to be inactive. One of the highest is Mount Haddington (1500 metres) on James Ross Island, its form smoothed by several hundred metres of ice cover and silhouetted against the setting sun. If this or a similar volcano were to erupt, large volumes of meltwater would be generated.

Use of glacier-mantled volcanoes in environmental reconstruction

Glacier fluctuations represent important stages in Earth's climatic evolution, but all too often the resulting deposits are difficult to date. In contrast, volcanic rocks are highly amenable to radiometric dating. So when glacial deposits and volcanic rocks are juxtaposed there is tremendous scope for obtaining a detailed record of environmental change. Where such studies have been undertaken, as in Iceland or on James Ross Island, scientists have found that glaciations go back several million years.

10 Shaping the landscape

The effects of glacial erosion and deposition are abundantly clear in many of the world's most beautiful mountain regions. Although now devoid of glaciers, regions such as the greater part of the Rocky Mountains of the USA or the uplands of the British Isles provide a rich legacy of glacial phenomena. Sharp peaks and steep-sided, flat-bottomed valleys are typical manifestations of the erosive capability of glaciers, while deposition from glacier ice has produced a variety of heaps of sand, gravel and mixed sediments. Even more abundant glacial deposits are found in the lowland regions bordering mountain ranges, or on the plains that underlie the outer limits of the last great ice sheets.

Glacial erosional landforms

The effects of erosion can be seen on all scales – from small outcrops of bedrock to the world's highest peaks, and to the vast areas of scoured low rocky country of the Canadian Shield. The distinctive imprint left by glaciers permits us to recognize the effects of glaciation in areas that have not been covered by ice for many thousands or even millions of years.

Small-scale features

On the smallest scale of centimetres to metres, glacial erosion is represented by striated, polished and grooved rock surfaces, features which are the result of debris-laden ice at the base of the glacier sliding over a slab of rock. Associated with them are smaller features such as **chattermarks** and **crescentic gouges** which are the result of a repeated juddering of an ice-embedded stone on the rock surface. In addition, the glacier can 'pluck' at the bedrock, creating a jagged rock-face, especially downstream of a bump in the bed.

The dramatic granite peaks of the Torres del Paine tower above a moraine-dammed lake and the person standing amongst the huge blocks that make up the moraine. A glacier filled the whole lake basin less than 200 years ago, and the only remnants occur on the intermediate bench above the lake and as a debris-covered calving mass at the foot of the cliff to the right.

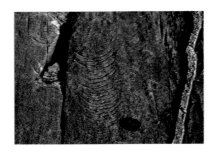

Crescent-shaped chattermarks and striations on bedrock of gabbro, Isle of Skye, Scotland. These features are the result of the juddering effect of debris-bearing ice as it slides over the bedrock. Ice flow was from top to the bottom of the picture.

Roches moutonnées were used widely in Scandinavia by bronze age people. The smooth rock surfaces were an ideal base for petroglyphs often showing people, boats and animals. This spectacular scene is near the south shore of Lake Mälaren (close to Sundbyholm) was created much later during the early eleventh century. Locally known as 'Sigurdsristning', it shows the fight of Sigurd against the dragon.

Other features, eroded with the help of subglacial meltwater under pressure, take the form of small, irregular, smooth hollows in bedrock, known as plastically moulded forms or **p-forms**. Such features form by a combination of erosion by sediment-laden meltwater, wet slurry-like glacial sediment and direct glacial abrasion. Discrete channel forms, cut into solid bedrock, are referred to as **Nye channels** after the British physicist who first defined numerically how they formed.

Intermediate-scale features

The smaller features are often mirrored on a scale of tens to hundreds of metres. For example, large **grooves** in areas of scoured low relief, that are many times longer than they are wide, resemble striations in form. Other larger features include rocky knolls that are convex and smooth but striated on their upstream side, and irregular and plucked on the slope facing downstream. These knolls, which are known as **roches moutonnées** after a wavy French wig popular in the eighteenth century, are common at the bottom and

The combination of glacial abrasion (producing striae) and subglacial meltwater under high pressure gives rise to channel-like structures (called Nye channels) cut in the bedrock, as at the margin of Glacier de Tsanfleron, Switzerland. View away from glacier across limestone bedrock to the ranges in the Valais Alps (Weisshorn, right of centre).

Small-scale features of glacial erosion are striations (fine scratches) resulting from the abrasive characteristics of debris-laden ice sliding over bedrock. The whaleback form shown here is a roche moutonnée on the island of Sandham in the Stockholm archipelago. It is abraded on the upstream side to the left and was plucked by ice on the right.

The erosion of a mountain from all sides by glaciers produces a 'horn' with steep arêtes rising to a sharp peak. Ama Dablam 6856 metres) viewed from the Khumbu Valley is one of the world's finest horns.

along the flanks of glaciated valleys. A large fine example is the Lembert Dome in Yosemite National Park, California. Sometimes deposition occurs in the lee of a rock outcrop, and the result is a **crag-and-tail** feature, such as the classic example of Edinburgh and the Royal Mile in Scotland. These large-scale landforms are described as follows.

The Lembert Dome in Yosemite National Park, California is an exceptionally large roche moutonnée. It clearly demonstrates ice flow from right to left, the abraded slope providing an easy walk to the top, while the plucked craggy slope to the left is a challenging rock climb.

Below. A sunset over the Cordillera Huayhuash in Peru brings out the intricate detail of the arête on Nevado Jirishanca (6019 metres).

A series of cirques on the north flank of the Glyderau range in Snowdonia, Wales, viewed from Tryfan. The centre left peak is Y Garn (943 metres), and the tarn below is Llyn Idwal. It was in the cirque containing this tarn that Charles Darwin recognized evidence of glaciation for the first time in Wales, in 1842.

Large-scale features

The larger-scale landforms in mountain regions are the most impressive features of glacial activity, as in the glacial erosional landscapes of the Scottish Highlands, the English Lake District and Snowdonia in Wales; parts of the Rocky Mountains and Sierra Nevada in North America; the Pyrenees, much of Scandinavia, and the Alps in continental Europe; and the Urals and ranges in Siberia and Japan in Asia. In low latitudes, abundant forms occur in the Himalayan and associated mountains, and even near the summits of the highest mountains of Africa and New Guinea. Straddling both hemispheres, the Andes also have large-scale glacial landforms, as do New Zealand's Southern Alps and the Tasmanian highlands in Australia.

Cirques, arêtes and horns

In mountain areas without extreme contrasts of height, rocky hollows can often be found that have been excavated out of the higher parts of the mountains by small glaciers that have eroded backwards and downwards. These hollows are known internation-ally as **cirques** (from the French), although terms like **corrie** (Scottish) and **cwm** (Welsh) are also widely used in Britain. Cirques typically measure a few kilometres in length and width, and about half to a third of their length in height. However some, such as the Walcott Cirque in Antarctica, measure tens of kilometres in length. But whether it is a large one, a moderate-sized one like the Western Cwm on Mount Everest, or a small one such as is found in the English Lake District, the length-to-height ratio is much the same. All cirques have a steep, commonly near-vertical headwall, and many have an over-deepened basin containing a small lake or **tarn**. In the steepest alpine terrain, however, headwall and downward erosion has normally not been sufficient to create such lakes and the cirque floors slope outwards.

Lauterbrunnental in the Berner Oberland of Switzerland. This valley is often regarded as a typical glaciated valley, in having a near U-shaped form in cross-section. True U-shapes such as this are relatively rare, however. The extremely steep valley sides result in numerous high waterfalls, such as Staubbachfall seen here above the village of Lauterbrunnen.

If cirques on opposite sides of a mountain erode backwards sufficiently they may meet and a steep-sided rock ridge, known as an **arête** (from the French) or **Grat** (German), develops. In Britain arêtes such as Crib Goch in Wales, Striding Edge in the Lake District, Aonach Eagach and the Cuillin Ridge in Scotland provide popular scrambling or climbing routes, and in the Alps or western Cordillera of North America, they provide aesthetically pleasing and challenging routes to the summits. Many of the first ascents of Alpine peaks, such as the Matterhorn and Weisshorn, were made by way of arêtes.

Where three or more cirque glaciers have eroded backwards to the extent that the highest ground is no longer immune from erosion, a sharp, pointed peak or **horn** is produced, with a series of arêtes between each pair of glaciers rising steeply up to the summit. The best known is the Matterhorn on the Swiss-Italian border. Other well-known and well-formed horns are Ama Dablam in Nepal,

Glencoe in the Grampian Highlands of Scotland is one of the finest glaciated valleys in Britain. This view down-valley from 'The Study' illustrates the parabolic curved cross-section that is more typical of glaciated valleys than a U-shape. The crags on the left are known as the Three Sisters, and represent spurs truncated by the valley glacier.

Opposite. Near-vertical walls, eroded by glaciers, are characteristic of some glaciated valleys. El Capitan in Yosemite Valley, California is a wall of granite, and represents one of the most famous climbing grounds in North America.

Glaciated valleys over-deepened below sea level are described as fjords. Spectacular examples, exceeding over 100 kilometres in length, are a feature of the coastline of East Greenland. Kejser Franz Josef Fjord is one of the longest, and is bounded by mountains approaching 2000 metres high, whilst the waters of the fjord are over 1000 metres deep in places.

Opposite. Mitre Peak (1692 metres) is a horn located in the inner reaches of Milford Sound, a popular tourist destination in Fiordland National Park, New Zealand. This view seawards also illustrates a hanging valley to the left of the peak.

Mount Assiniboine in the Canadian Rockies, Mount Aspiring in New Zealand's Southern Alps and K2 in the Karakorum Mountains of Pakistan. In Britain, one of the best examples is Cir Mhor on the Isle of Arran.

Glaciated valleys

Glaciated valleys that were formerly occupied by the principal glaciers descending from the higher mountains have a characteristic form. In cross-section they are often described as U-shaped, but only rarely are their sides truly vertical, and the term 'parabolic' is more appropriate. Lauterbrunnental in the Berner Oberland of Switzerland and the Yosemite Valley in California are excellent examples of valleys that can truly be regarded as U-shaped, the vertical walls providing rock climbers with extremely challenging ascent routes. In contrast, the more typical parabolic form is exemplified by Glencoe

Areal scouring is the product of powerful glacial erosion over relatively wide areas of subdued relief. Irregularities in bedrock structure give rise to the intricate detail of hillocks and lakes in the landscape. This example of areal scouring shows ancient crystalline rocks (with black dykes) in the Vestfold Hills of East Antarctica, looking towards the ice sheet from near the coast.

in the Grampian Highlands of Scotland, as well as by the many other glaciated valleys through the British Isles.

In many glaciated valleys enhanced erosion has created a series of rock basins, filled by lakes that are formed when a tributary glacier joins the main one. Later these lakes will be partly or totally filled with sediment. Many glacial valleys have a stepped longitudinal profile, or a series of rock barriers called **riegels** (a term derived from the German) that extend part or the whole way across the valley.

The level to which a valley glacier has eroded is clearly marked by a change in slope, as well as by the preservation of spurs above the glacier level. Up as far as the erosion level, the side of a glaciated valley normally consists of a single clean sweep, the spurs that would have existed in preglacial times having been truncated. Above the erosion level, the preglacial forms and deeply weathered bedrock may survive. Some tributary glaciers may not have had the erosive

One of the best known glacial meltwater channels in Britain is Newtondale, cut by overflow from ice-dammed Lake Pickering during the last glaciation. It carries a small stream that is out of proportion to the size of the valley. The North York Moors Railway, which uses steam locomotives to carry tourists, follows this route.

power to cut down to the level of the main valley, and hanging valleys result, often with fine waterfalls plunging from them. In other cases, valley glaciers may have spilled over the low saddles or cols of the bounding ridges into another valley, resulting in **breached watersheds**.

Fjords

Fjords are flooded over-deepened glaciated valleys, formed because valley glaciers erode deeply in their middle reaches and, if extending into the sea, to a depth well below sea level. The term 'fjord' is of Norwegian origin (spelt 'fiord' in North America), and it is Norway that has one of the finest fjord coastlines in the world. Almost its entire coast is indented with fjords, the longest and deepest being Sognefjord (200 kilometres long and 1300 metres deep). Many other moderate- to high-latitude countries have fjords too. Those in western Scotland are known as **sea lochs**, but are small by world standards. In contrast, Greenland, where fjords are still being formed, has the world's longest, with the combined Nordvestfjord and Scoresby Sund on the east coast measuring 350 kilometres in length. In the Americas well-developed fjord coastlines occur in southern Alaska, British Columbia and southern Chile, many of them still under the influence of glaciers. The islands of the High-Arctic have many smaller fjords, while on the other side of the world south-western New Zealand has a well-developed fjord coastline as impressive as any, and the Antarctic Peninsula has many fjords that are filled by glaciers. The sub-Antarctic island of South Georgia has numerous short fjords still occupied by glaciers.

Like glaciated valleys, fjords may comprise several rock basins, but the simplest of them are deepest at their heads and become gradually shallower towards the sea. Many fjords have a shallow or even partly exposed sill at their seaward limits, and some have near-vertical rock walls. Like valleys, they have U-shaped or parabolic cross-sections and, with time, they too become filled by sediment. **Hanging valleys** and waterfalls are common features along the flanks of fjords.

Scoured bedrock of low relief

The vast areas of the gently undulating Canadian Shield, parts of the Baltic Shield and the margins of the Greenland and Antarctic ice sheets contain large-scale glacial erosional landforms of a scenically less dramatic nature. The hard crystalline Precambrian rock that underlies these regions bears evidence of erosion along the structural grain of the bedrock, with long linear erosional forms. In many cases elongated lakes or boggy hollows were eroded out parallel to the bedrock structure. Such areas are regarded as having undergone **areal scouring**. Parts of northwest Scotland show the same landscape attributes, except that the scouring process is incomplete. Here, isolated sandstone peaks stand proud above the heavily scoured ancient crystalline rocks with their little knolls and small lakes; this is called **knock-and-lochan topography**.

Erratic boulders, such as this north of Zug on the Swiss Plateau transported by the ice age Reussgletscher, were first used by scientists such as Louis Agassiz in the early nineteenth century to hypothesize that glaciers were once much more extensive.

Meltwater channels

A variety of channel-like forms, typically tens of kilometres long and a kilometre or two wide, are a feature of glacial landscapes formed towards the end of the last glaciation when meltwater was abundant. Some channels have the typical downhill profile of any normal river valley, and are viewed as **overspill channels**, where glacial lakes have escaped over a low point in a barrier. A good example in Britain is the overspill from former Lake Pickering in Yorkshire. Other meltwater channels have 'up-and-over' profiles that must have formed sub-glacially. In such situations water is under pressure and is able to flow uphill if there is a sufficient head of water. Northern England and Wales are well endowed with such features, but networks and individual channels may be found in almost any glaciated landscape.

Glacial depositional landforms

It is, perhaps, appropriate to start this section with a quotation from the *Bible*, as this underpinned much of the early thinking about how the extensive unconsolidated deposits of Europe were formed.

The lower slopes of Val d'Hèrens in Canton Valais, Switzerland are draped with thick deposits of 'till' – sediment released directly from a glacier. Near the village of Euseigne erosion of the till has left these well-consolidated pinnacles, capped by boulders, through which engineers have driven the main road up the valley.

Till exposed in a roadside quarry above the north shore of Loch Torridon in the Northwest Highlands of Scotland. The texture is clearly visible with large boulders ranging down to clay-size material. The reddish colour is from the ancient sandstones that are a feature of this region.

> Towards the end of seven days the waters of the flood came upon the earth. In the year when Noah was six hundred years old, on the seventeenth day of the second month, on that very day, all the springs of the great abyss broke through, the windows of the sky were opened, and rain fell on the earth for forty days and forty nights. . . . More and more the waters increased over the earth until they covered all the high mountains everywhere under heaven. The waters increased and the mountains were covered to a depth of fifteen cubits. Every living creature that moves on the earth perished, birds, cattle, wild animals, all reptiles, and all mankind . . . only Noah and his company in the ark survived
>
> (Genesis 7: 10–12, 20–21, 23).

This was the explanation widely held in the early nineteenth century to explain the distribution over northern Europe of so-called **drift** deposits – unconsolidated sediments which we now know to be of glacial and glaciofluvial origin. At the time controversy raged as to whether or not the *Bible* should be taken literally, and this particularly affected geologists. It was recognized that many large blocks of rock, now called **erratics**, scattered over much of northern Europe had been transported great distances, for example from the

Although the deposits released by a glacier (till) have a wide mixture of particle sizes, meltwater streams rework and sort this sediment, producing sand and gravel. This example of sand at Banc-y-Warren near Cardigan, Wales shows faults that were generated as ice around the deposit melted. Such deposits are a valuable resource for the construction industry.

Glaciers carry large amounts of mud that is also removed by meltwater streams, and then collects in hollows. When it dries out, 'desiccation cracks' form, as here near Kongsvegen in northwest Spitsbergen. The polar bear footprints (about 30 centimetres in diameter) give the scale. We met the culprit about half an hour later.

Scandinavian mountains all the way to Denmark. The influence of glaciers was not then known, and the geologists of the day could only conceive that the blocks were moved during catastrophic floods. Thus Noah's Flood was the widely accepted explanation for these deposits, which were given the name *Diluvium* (the Latin word for flood or deluge).

In fact, the idea that glaciers could be responsible for transporting large boulders was already gaining currency, following the work of Swiss naturalists, notably Louis Agassiz in the Alps in the late eighteenth and early nineteenth centuries, but the widespread acceptance and application of the theory only grew very slowly.

Glacial and meltwater deposits are the components of a wide range of interesting landforms, which, although less dramatic than erosional ones, are nevertheless distinctive features of the environment. The rolling countryside with fertile fields and wooded knolls, so characteristic of the lowlands of central and northern Europe and the northern Mid-West of the USA, are dominated by such features. Yet, although best developed in the lowland areas that once were covered by continental ice sheets, depositional landforms are also present in highland areas where glacial erosion was dominant.

When glaciers advance they push large amounts of debris before them. Then, as they recede, large ridges or moraines of loose debris are left behind. In this example, lateral moraines join with a terminal moraine to enclose a lake in front of Imja Glacier in the Khumbu Himal, Nepal. Rapid recession has exposed a steep unstable inner face that occasionally collapses into the lake, whilst fine sand and silt is caught up in the wind and redeposited elsewhere. The collapse of such moraines is of concern in high mountain regions, because of the potential for lake-outburst floods.

Moraines

The most easily identifiable of depositional landforms are **moraines**, since they are commonly long, sharp-crested ridges or ridge complexes, made up of a mixture of till and other deposits pushed up during glacial advances. In a valley setting, the furthest advance of a glacier is marked by a **terminal moraine**, normally arcuate in form, reflecting the original shape of the glacier snout. Such features usually range from a few to 50 or more metres in height but, because powerful streams normally accompany glacier recession, only remnants may remain. In lowland areas the advances of ice sheets created even larger ridge complexes that can sometimes be traced for hundreds of kilometres, which were breached in places by huge streams issuing from the receding ice. Lakes frequently formed behind these terminal moraine complexes, the Great Lakes of North America and Zürichsee in Switzerland being prominent examples.

Twelve-thousand-year-old hummocky moraines in Coire a' Cheud Cnoic (Corrie of a hundred hills), Glen Torridon, northwest Scotland are amongst the best preserved glacial depositional features in Great Britain. Ice flow was from top right to mid-left. The small white cottage in front of the moraines gives the scale.

The clean white snout of Matanuska Glacier in Alaska contrasts strongly with the grey debris in front of it, most of which was released from the base of the glacier. Remnants of glacier become buried by debris and, as the ice slowly melts, small hollows called kettle holes form, filling with ponds. Vegetation is rapidly taking hold on the mineral-rich sediment in the proglacial area of this glacier and will soon become dense forest.

In some areas terminal moraines are not simply dumped at the ice margin, but are the product of deformation (notably thrusting), both at the ice margin and beyond. Propagation of deformation from a glacier into the forefield means that the terminal moraine may actually form a few hundred metres from the ice front.

Lateral moraines are evident along the sides of the valleys, but these have a much poorer chance of surviving; hill-slope movement and subsequent rock-falls combine to obliterate them. However, in areas of recent recession, unstable ridges dating back to the 'Little Ice Age' of around 1650–1850 are still very much in evidence high above the present glacier tongue and down-valley of the glacier's snout. The ridge commonly stands out from the valley sides, creating a so-called **ablation valley**, with its own stream and ponds, between the ridge and the valley side, well above the floor of the main valley. A steep face of bouldery till, held temporarily together by the clay in the deposit, is a feature formed on the glacier-facing side of such moraines, and today rainwater creates a furrowed pattern down them, making scrambling up or down them surprisingly difficult and dangerous. In contrast, the side of the moraine nearer the hillside tends to be well covered, plant growth having been possible even when the ice was in position. Once the ice has gone, these ridges collapse rapidly.

Where a valley opens out, lateral moraines may leave the valley sides and merge with terminal moraines. Between the moraines and the receding ice margin, an assemblage of hillocks and hollows, filled with water, is present. This zone includes **hummocky moraines** that traditionally have been regarded as the product of ice stagnation. Some hummocky moraines show alignment parallel to the ice margin and are the product of active recession and deformation, notably by thrusting, in the ice. Hummocky moraines of fresh-looking appearance are found in those parts of upland Britain that were affected by the last pulse of glaciation about 10 000 years ago. Other examples were associated with the recession of the Laurentide Ice Sheet, with well-known examples associated with the Des Moines lobe in the American Mid-West.

Each of the above types of moraine when first formed may have a core of ice, and the volume of debris may be small; dark wet patches of unstable debris are clear indications of an ice core. These ice-cored moraines may survive for many decades or even centuries, especially in the Polar Regions.

There are other types of moraine up to a few metres high associated with existing glaciers, but they rarely survive for more than a few decades. Small **annual push moraines** may form as a glacier advances a few metres in winter when ablation largely ceases, and a whole series of such parallel ridges may develop if the longer-term trend is one of recession. Sometimes, ice movement across a plain of till may

Meltwater commonly flows in channels beneath glaciers. They cut upwards into the ice, and then may become clogged with fluvial sediment. When the ice recedes an upstanding, sinuous, flat-topped ridge of sand and gravel may be left behind. This example is at the tidewater glacier of Comfortlessbreen in northwest Spitsbergen.

generate **fluted moraine** – long, straight, parallel, smoothly rounded ridges parallel to the ice-flow direction, which may be exposed if the ice front recedes steadily without in situ stagnation.

A special set of moraines is associated with surge-type glaciers that periodically advance catastrophically, as described in Chapter 5. Debris may be thrust up from the glacier bed during a surge to form curving ridges a few metres high parallel to the snout. At the end of the surge wholesale stagnation of ice occurs, creating a hummocky and pitted morainic topography on which the thrust ridges have been let down. Discriminating between non-surge and surge-type moraines is, in fact, an important way of assessing whether a particular glacier advance was the result of either climatic change or periodic instability of the ice mass.

If a glacier or ice sheet rides over a plain of till or heaps of moraines, the material becomes quite gooey and is easily moulded into new shapes or eroded. One product of this process is the generation of 'drumlins', streamlined hillocks of till, sometimes draped over bedrock knolls. A drumlin is shaped like an inverted spoon, with the steep slope in the upstream direction, and is orientated parallel to the ice-flow direction. They reach 100 metres or more in length and are up to 50 metres high, often covering large areas in a type of landscape known as basket-of-eggs topography. The Eden Valley of north-west England is a good example, and extensive drumlin fields also occur in New England in the USA and in Northern Ireland.

Although glacial transport is best demonstrated by the above landforms, isolated erratics are also good indicators of the former extent of ice and the directions of flow. Large blocks that fell on a glacier surface, or boulders in the bed of a glacier may end up hundreds of kilometres from their place of origin.

In some cases, moraines may contain valuable minerals, and even if the moraine itself is uneconomic, by reconstructing flow paths back to the source rocks it is possible to locate viable mineral reserves. This method of mineral location has been used most widely in Canada and Finland.

Glacial meltwater as an agent of deposition

Meltwater within, at the side, or beyond the limits of a glacier also creates landforms. Wide fluctuations in discharge between summer and winter create unstable 'braided' stream-channel systems, where the streams modify, sort and redistribute glacial debris creating **outwash plains**, both beyond the terminal moraine system and directly in front of the receding glacier. Such systems normally extend across the entire width of a flat-bottomed valley, as can be seen in some fine examples in New Zealand and Alaska. In lowland areas these plains may be many tens of kilometres wide. They are sometimes known as **sandar** (singular **sandur**), a term from Iceland where they extend along much of the south coast, in the strip of land between the ice cap of Vatnajökull and the sea.

Tasman River with Mount Cook in the background is a typical braided river, in this case emanating from the Mueller and Hooker glaciers to the left and the Tasman Glacier to the right. Braided rivers are characterized by massive variations in discharge, so channel switching is common, and vegetation has difficulty becoming established.

Glacial outwash deposits often bury remnants of dead glacier ice. As the ice slowly melts, steep-sided, water-filled hollows, known as kettles or **kettle holes** (a term of Celtic origin), develop. Beneath the glacier a stream cuts a channel upwards into the ice, and may eventually become choked with debris. As the ice melts, long, narrow ridges of sand and gravel may be left standing above the general level of the outwash plain, in features known as **eskers** (another Celtic term). Varying in height from a few to tens of metres, they wind for hundreds of metres across the landscape. In Finland, which is particularly noted for them, they provide convenient flat-topped ridges on which roads have been built through lake-studded country.

Stream deposits frequently accumulate adjacent to the ice, which are left as isolated hillocks when the ice melts. The term **kame** (an old Scottish word) is used to describe ice-contact meltwater deposits formed in this way parallel to the ice front, and **kame terraces** are formed at the side of a glacier, especially where a tributary stream enters the main valley. As the ice melts, a level, gently sloping platform is left, perched on the hillside. In addition to stream deposits, a kame and kame terrace may also be composed of lake sediments.

Apart from their character as distinctive elements of a glacially influenced landscape, outwash, esker and kame deposits provide much of the sand and gravel needed for road and building construction around the temperate regions of the world, especially in Europe and North America.

Offshore depositional features

Many glaciers and ice sheets during the Ice Age advanced across high- and mid-latitude continental shelves. By using sophisticated geophysical techniques such as seismic profiling and side-scan sonar, scientists have identified many of the same depositional features on the sea floor, much as they occur on land; flutes and moraines are examples. In addition, there are large-scale depositional forms known as **trough-mouth fans**, formed where large amounts of sediment have been delivered to the edge of the continental shelf by ice streams and dumped over the edge. Trough-mouth fans are well-studied off the coast of Norway, on the western edge of the Barents Shelf and in Prydz Bay in Antarctica. Some of them measure more than 100 kilometres across.

Glacial landscapes are the product not just of glacial erosion and deposition, but also of associated processes such as running water, wind and mass movement. It is no wonder then that for the scientist to understand the processes responsible for such landscapes, considerable detailed research is necessary. This chapter thus provides only the briefest of summaries of the components of the glacial landscape.

11 Glaciers and wildlife

Surprising as it may seem, glaciers and their surroundings are often havens for wildlife. Camping on a glacier a thousand metres up amongst the icefields of Spitsbergen the authors have been disturbed at night by the constant chatter and chuckling of a colony of fulmars, nesting on bare rock cliffs nearby, despite being 30 kilometres from the coast. We have heard the cry of an Arctic fox, and looked out to see that elegant, white-coated animal trotting back and forth below the cliffs, eyeing the nesting birds with eager anticipation, and obviously waiting for an unfortunate chick to fall out of its nest.

Glaciers are not totally lifeless, despite the harshness of the environment. All around the world, many different animal and plant species live and die on and around them, uniquely adapted to the cold. Some species are confined to the edges of the polar ice sheets while others make their homes around mountain glaciers.

Antarctica

The most hostile environment on Earth is Antarctica. The entire continent is a cold desert, characterized by low snowfall, lack of water, exposure to the wind, and salt-bearing mineral soils that lack organic matter. The number of species capable of living under these adverse conditions is small. Only twelve species of birds and four of seals breed in the Antarctic. However, the ecologically rich seas bordering the continent teem with life. The low diversity of species reflects the continent's isolation from the nearest land; it is separated by 650 kilometres of stormy seas that discourage migration.

Land animals in Antarctica are small; indeed, there is nothing bigger than a horsefly, with the mite being the most common creature living entirely on the land. In contrast, the sea is home to many

Emperor penguins (*Aplendytes forsteri*) at the edge of the fast ice in McMurdo Sound in spring. They are well known for breeding during the winter in Antarctica, and this group's rookery lies just off the northern tip of glacierized Ross Island in the background.

Adélie penguin and chick (*Pygoscelis adeliae*) near Casey Station, east Antarctica. Rocky sites, in the narrow strip of land between the ice sheet and the sea, provide ideal nesting sites, especially when there is a supply of glacially transported stones to construct the nest.

large, warm-blooded creatures that are top of the food chain; for example, the world's largest seabirds (such as the albatross and emperor penguin), seals and whales. One of its residents, the blue whale, is the largest animal the world has ever known, growing fat on the abundance of food, especially krill, that the sea provides. Unfortunately, humans have ruthlessly hunted this beast to the extent that it now needs total protection, yet even so recovery is slow.

Most marine birds and mammals spend more of their time at sea than on the ice, and feed entirely in the water. However, some sea birds roost and breed far inland. For example, snow petrels and Antarctic skuas have been found in small colonies 2000 metres high on the nunataks of Dronning Maud Land, 300 kilometres from the sea. Skuas venture even further into the icy interior of Antarctica, and the ill-fated polar party of Captain R.F. Scott in 1912 recorded one bird a mere 250 kilometres from the South Pole, more than 1200 kilometres from the nearest open water.

To many people, the penguin holds a particular fascination. Two species, the emperor and the Adélie, breed on the continent, especially on stable sea ice, or on ledges and caves in ice cliffs and ice shelves, and even more breed on the more northerly offshore islands. More than any other bird, the emperor penguin is uniquely adapted to the frigid climate, although its population is estimated to be only about a quarter of a million. It lays its eggs in early winter and incubates them through the coldest months of the year. The males take over from the females shortly after the eggs have been laid, huddling together against the fierce blizzards to conserve heat, with the eggs on their feet, covered by a fold of feathered skin. After about 65 days, at about the time that hatching takes place, the now well-fed females return, allowing the hungry male to plod north across the sea ice to find food in open water.

The emperor's breeding habits have long been a source of fascination to biologists, ever since Edward Wilson, chief scientist on Scott's last expedition, together with 'Birdie' Bowers and Apsley Cherry-Garrard in 1911 made one of the most arduous journeys ever undertaken in the cause of science. Graphically described by Cherry-

Elephant seal (*Mirounga leonine*) on the sandy beach at Davis Station (East Antarctica). This is one of several dozen animals that come ashore each spring to moult and gather together in a large heap.

Garrard in *The Worst Journey in the World* (published in 1922 by Constable, London), this expedition was undertaken in the darkness of deepest winter in order to capture an emperor penguin egg soon after it was laid. The explorers hoped that the primitive embryo would establish a link between reptilian scales and feathers, and thus throw light on the origin of all birds. The terrible privations this party suffered give a good indication of the conditions experienced by those emperor penguins that remain at the glacier edge all winter. Pulling sledges weighing 343 kilograms, the three men had to cross part of the Ross Ice Shelf, where the temperature plummeted to $-61\,°C$ ($-77\,°F$) and often failed to reach $-50\,°C$ ($-60\,°F$) all day. Eventually, the party reached Cape Crozier and, after a difficult scramble down icy slopes, reached the penguin colony. After col-

Weddell seal mother and large pup (*Leptonchotes weddelli*) resting on sea ice near the glacier-fringed coast of Granite Harbour, Ross Sea, Antarctica.

Antarctic skua (*Catharacta antarctica*) at Mirnyy Station, East Antarctica. This large aggressive bird predates on Adélie penguin chicks and eggs, but is also known to journey far south across the ice sheet, following sledging parties. It will also scavenge around human camps and stations, if food or rubbish are not disposed of properly.

lecting a number of eggs, the men established a camp above the colony, only to have it swept away during a violent blizzard. For three days they huddled under a groundsheet ensconced in their iced-up reindeer skin sleeping bags, while the blizzard shrieked over their heads. Fortunately, they found the tent undamaged, for without it their survival would have been in doubt. Exhausted by the hardships of the preceding four weeks, the party now had to retrace its steps to the expedition base, nearly 100 kilometres away. Eventually they made it back after 36 days, after enduring probably the harshest conditions ever to be survived by any men. Sadly, both Wilson and Bowers lost their lives later in the expedition, after reaching the South Pole with Scott.

Penguins and seals probably spend half their lives in the waters surrounding Antarctica, which for most of the year are warmer than the air. Warm-blooded animals are well adapted to take advantage of the rich food resources in the sea. Heat loss is reduced to a minimum by dense, water-repellent plumage in the penguin and petrel, and by

Arctic hare (*Lepus arcticus*) on sparsely vegetated tundra near Thompson Glacier, Axel Heiberg Island, Canadian Arctic.

the thick, tough skin of the whale and seal, and a thick layer of subcutaneous fat or blubber provides additional insulation. A further adaptive feature is their compactness, the extremities of seals and penguins are generally short and bony, so little blood needs to circulate through exposed areas. Their large size is also an advantage in the cold climate, since a low ratio of surface area to volume conserves heat more efficiently than the large ratio of a small animal.

Although seals and birds can tolerate very low temperatures in still air, they prefer to be in water when it is windy. However, if they have to, such as when they are protecting their young, they are capable of remaining at their post without harm in blizzards that last several days. The young may not be so fortunate, and in severe storms the mortality amongst juveniles is high.

Some animals do curious things in Antarctica. The strangest is the behaviour of crabeater seals in crawling inland over rocky debris or up glaciers for many kilometres to die. Dried out carcasses have been seen as much as 750 metres above sea level and 70 kilometres from the coast, and one of the authors has seen a desiccated seal melting out from the snout of Suess Glacier in the Dry Valleys.

Antarctica had a lush cover of vegetation, including southern beech, until the ice sheet developed about 36 million years ago. Forests and then tundra-type vegetation co-existed with the early glaciers until dramatic cooling produced the cold arid conditions that prevail today. However, even today Antarctica is not as devoid of vegetation as might appear at a first glance. There are various species of algae, lichens, mosses and fungi. The greatest diversity occurs in the warmer maritime parts of the Antarctic Peninsula and offshore islands, where even a couple of species of flowering plants occur. Mosses and lichens have been recorded on nunataks as far south as latitudes 84° and 86° respectively.

When ice recedes the first plants to colonize are algae and soil micro-organisms, with mosses and lichens appearing later, as can be observed both on morainic debris and bare rocks. They all grow best along ephemeral water-courses and around snow-banks. Antarctic plants survive because they have developed a strong resistance to frost and drying out, and have acquired the ability to grow rapidly in the brief periods when conditions are favourable.

The Arctic

The polar desert environment that characterizes much of Antarctica is found in a few inland parts of the High-Arctic, but is less impoverished. Indeed, most parts of the Arctic have a much more tolerable climate, so biologically the region differs greatly from Antarctica. Although extensive glacierized areas do exist in the Arctic, ice-free areas are much more widespread. Furthermore, parts of the Arctic remained ice-free even during the last Ice Age, allowing evolution to proceed without interruption. Thus, there is a much greater diversity of plant and animal species than in the South Polar Region.

The most striking difference is the presence of large land mammals that roam freely across snow and icefields as well as the tundra – the sparsely vegetated ground that is underlain by a thick layer of permanently frozen ground called permafrost. A total of 48 species of land mammals are found in the Arctic, comprising shrews, hares, rodents, wolves, foxes, bears and deer, but, although more than Antarctica, even this diversity is low compared with more temperate regions. However, only a few of these live in proximity to the glaciers. Greenland, for example, has only nine species: Arctic hare, Arctic lemming, grey wolf, Arctic fox, polar bear, ermine, wolverine, caribou and musk ox.

Arctic animals illustrate a variety of adaptations to the hostile winter conditions. Some avoid the severity of winter by migrating

Arctic wolf (*Canis lupus*), a regular visitor to a glaciological field station on Axel Heiberg Island, Canadian Arctic, trotting beside Colour Lake beneath Wolf Mountain.

south; for example, the majority of birds, and many of the herds of caribou in North America. On the other hand, the reindeer on Svalbard have nowhere to go, and seem to thrive on the cold: you often see them in summer cooling off on glaciers and snow patches. Other animals go into hibernation, such as the Alaskan marmot and Arctic ground squirrel. Although bears strictly do not hibernate, the grizzly, which is found around the glaciers of Alaska, enjoys a long period of sleep in winter.

The Arctic fox is adapted in several interesting ways. It remains active all the year round, changing its coat in autumn brownish grey to dense white to match the snow of winter, and has furry paws to insulate its feet from the frozen ground. It is able to maintain a steady heat production, and only needs to increase it when the tem-

Svalbard reindeer (*Rangifer tarundus platyrhynchus*) crossing snow-covered tundra against a backdrop of a frozen fjord and glacierized peaks on the peninsular of Brøggerhalvøya, northwest Spitsbergen.

perature declines to −40 °C. It ranges widely through the Arctic, not only venturing on to glaciers as mentioned above, but also onto sea ice. Indeed it has been sighted within 140 kilometres of the North Pole, 80 kilometres from the nearest land. The fox lives off birds and lemmings, and scavenges the remains of seals killed by polar bears. It can handle quite large birds. We have observed a fox den in a moraine close to a glacier in northwestern Spitsbergen. With two well-grown cubs playing outside the den, the mother went off to the neighbouring settlement of Ny-Ålesund to raid the nesting site of a flock of barnacle geese. To carry a two-kilogram bird, at least half its own weight, for five kilometres back to its cubs is an impressive feat. The fox not only killed the birds, but also ate and destroyed the eggs so systematically that only a handful of geese hatched any chicks that season.

Arctic fox (*Alopex lagopus*) in summer coat near the glacier midre Lovénbreen, northwest Spitsbergen.

Musk oxen (*Ovibos muschatus*) in defensive formation near Thompson Glacier, Axel Heiberg Island, Canadian Arctic.

The carnivorous grey wolf is a survivor from the Ice Age that spread through much of the Arctic when the ice receded. Hunting in packs, it feeds principally off sick and weak caribou in Canada and Alaska, but in Greenland relies on small animals. One of the authors lived alongside two wolves, a male and a female, during a three-month field season on Axel Heiberg Island. The male showed great interest in the various human activities in what it clearly considered its own territory. It often followed us around, frequently making clear its territorial claims in typical canine fashion. Interestingly, it was not food which kept the wolf close to us – it never accepted any human offerings. For this author, there is nothing more memorable of life in the Arctic than listening to the wolf's melodious howling under the midnight Sun, or on waking up looking out of the tent directly into the face of a friendly Arctic wolf.

Like the fox and wolf, the musk ox remains active throughout the year, favouring dry areas where the snowfall is low, so that it can scrape away at mosses and lichens. It frequents the tundra around the glaciers of the Canadian Arctic Archipelago and Greenland. The musk ox carries a thick, winter coat that is softer than normal wool. In summer the animal moults, and acquires a very untidy appearance, as it gradually loses stringers of knotted wool. When threatened with danger, family groups of musk oxen arrange themselves in a defensive line or circle, with the calves pushed to the back. This form of defence was very effective against predators such as the wolf, but was responsible for its undoing when rifle-bearing hunters began to decimate them. Glaciologists have reported several unfriendly encounters with musk oxen, usually old males that have been pushed out of the herd. Once in East Greenland, one of the authors, making notes quietly on the tundra, was suddenly picked on by a solitary musk ox, and without warning charged. With a cliff behind, there was no escape route, so the only option was to face the charging animal and shout abuse at him! This proved effective, since about three metres away, the beast had second thoughts, braking in a cloud of dust before backing off slowly.

The Arctic hare differs from its southern cousins in having a much thicker insulation. In winter it turns white, and in more northern parts remains white all year round. Unlike most species of hare it sometimes forms large herds, but mostly can be seen in groups of half a dozen or so. Hares have been a considerable source of amusement for both authors during field seasons in the Canadian Arctic and Greenland. One of the most amusing things we saw was their reaction after dozing in the warm summer sun outside base camp. After being disturbed by us they would become startled and race away, strangely speeding around in circles not very far away from us. Another aspect of their character is to run as all hares do, but then go into a frenetic hopping mode, before continuing their run.

Small animals, such as rodents, cannot compensate for heat loss by improving their insulation, so instead they increase their metabolic rate. They also use the insulation provided by their surroundings – by sheltering beneath the snow cover. Lemmings, therefore, do not strictly hibernate, but nevertheless survive even on the northernmost arctic islands.

The world's largest carnivore, the polar bear, epitomizes the animal world's adaptation to the harsh Arctic environment. Polar bears are widely, but thinly, distributed over the Arctic sea ice, but

Ptarmigan (*Lagopus mutus hyperboreus*) in winter plumage, a year-round resident of the Arctic tundra near Ny-Ålesund, northwest Spitsbergen.

they breed in onshore dens, usually in snow-banks on barren coastal hillsides. They range over hundreds of kilometres, right across the frozen Arctic Ocean. Although they rarely venture onto glaciers, they have been known to cross icefields, such as on Svalbard, when rapid break up of the sea ice on the archipelago's west coast makes it impossible for the bear to catch seals, forcing it to cross the land ice to reach the more permanent sea ice on the east coast. It is important to avoid a confrontation with a polar bear because it can attack a human without warning. Shooting a bear in self-defence is a last resort, as the animal is a protected species, yet regulations across the Arctic require visitors to carry a rifle for protection. One of us had two encounters with bears in the same season in western Spitsbergen. On the first occasion, just after working on the cliff of the tidewater glacier Konsvegen, which encloses a narrow beach and a marine embayment, four of us were scrambling over coastal

The small polar willow shrub (*Salix polaris*) on Axel Heiberg Island, one of the earliest shrubs to colonize ground left bare by glaciers.

Purple saxifrage (*Saxifrage oppositifolia*), one of the earliest plants to colonize deglaciated terrain, Axel Heiberg Island.

White arctic bell heather (*Cassiope tetragona*), Kong Oscar Fjord, East Greenland.

moraines and unexpectedly met a large male bear coming towards us. Ducking behind the moraines we ran upslope across flowing debris flows of glacial sediment, hoping the bear had not seen us. Pausing for breath (and photographs) behind a hummock, we saw the bear continue plodding along the beach after looking unnervingly in our direction, and into the embayment where we had been working. A meeting there would have been serious, as there would have been no escape, and we may well have had to use the rifle. On the second occasion, some of us were camping outside a glaciological hut near a glacier named Finsterwalderbreen. It is customary to surround tents with trip wires with flares, the idea being that the bang would frighten the bear away if it came too close. As it happened one of the party had set the flares off several times accidentally, so when, early one morning, another went off, we thought

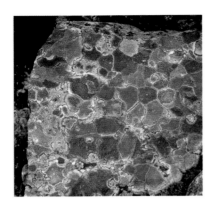

Lichens covering a glacially transported boulder in the Little Ice Age moraine (c. AD 1750) of Storglaciären, Kebnakaise, northern Sweden.

nothing of it, until an exclamation from one of our group of, 'There's a bear outside', followed by a rifle shot. Then we all looked out of our tents, another shot was fired and the bear (a mother) trotted off and joined her cub. It transpired that when our companion had looked out of his tent, he saw a snorting swaying bear just a couple of metres away. The flare had failed to scare the bear, so the first rifle shot, through the roof of the tent, was meant to scare the bear away. It failed to do this, but fortunately the second warning shot did so. After bear sightings this close, glaciological studies became somewhat tense, but on the latter occasion we had the fortune of watching a bear playing with her two cubs on ice floes just offshore.

Few species of bird remain on the Arctic tundra during the winter, and those that do, such as the ptarmigan, have adapted by reducing their metabolic rate at low temperatures and increasing the density of their feathers. However, in summer the Arctic supports nearly 200 species of bird, which have migrated both overland and over the sea. They arrive in springtime as the snow begins to melt, plants begin to flower and the days become long. One of the most impressive sights in nature is the thousands of seabirds nesting on, and wheeling about, the steep cliffs of a fjord, with a glacier calving at its head, producing small icebergs which provide a perch for the birds, and also for seals. Although most birds prefer to nest on coastal cliffs, some sea birds prefer nunataks amongst the highland icefields, so few parts of the Arctic are devoid of life. Some breeding birds cover enormous distances, but none can compete with the Arctic tern, which nests in the Arctic and migrates to the Antarctic for the southern hemisphere summer.

In terms of numbers, insects are the most abundant of Arctic creatures. They play a vital role in the ecosystem, providing an important source of food for many birds. Mosquitoes and midges are particularly noticeable because they inflict so much misery on travellers when the weather is warm, but there are many other less intrusive insect species.

Arctic plants have also adapted to the severe climate in several

Steinbock or stone ibex (*Ibex ibex*) are ideally adapted to life at high altitudes in the Alps. Comparatively large body mass and dense fur help this animal to maintain body heat during winter. In summer keeping cool can become a problem; so groups of Steinbock occasionally seek relief from summer heat on snow patches or glaciers.

ways. Most obvious is their smallness of size, which minimizes desiccation from winds and allows them to nestle under an insulating cover of snow. Temperatures are also higher at ground level during the growing season, facilitating flower and seed production. Among their numerous other adaptations, a striking one is the early development of buds, which allows them to flower at the earliest opportunity in the spring. Their dark-coloured leaves and stems are also noticeable; these allow greater heat absorption, especially if the leaf cover is dense. Examples include the heathers and saxifrages, the leaves of which protect the plant from abrasion by snow.

Arctic plants have to be further adapted to seasonal variations in light intensity, such as an ability to vary their rates of photosynthesis. They also must withstand total freeze-up of the ground for eight to ten months of the year, as well as summertime frosts.

Alpine regions

In the mountainous regions of more temperate latitudes glaciers frequently descend into zones where the climate is much less severe than in the Polar Regions. Thus, when the ice recedes, plants may colonize the newly exposed ground at an amazing speed, enjoying the fertility that is often provided by sediments rich in minerals. The so-called pioneer plants are the first to arrive, having been transported in seed form by the wind and birds and then germinating in little more than sand and mud. These early plants include saxifrages and grasses, and may often be found within two years of the ice receding. Decaying organic material from these pioneer plants then prepares the ground for more demanding species, such as the larger flowering plants, *Alpenrosen* and asters, as well as shrubs of alder and

Vegetation in moist, mild areas such as near Fox Glacier, South Island, New Zealand, can generate rapidly even on dead glacier ice. Here a moraine from the early twentieth century has a thick cover of trees, although the ground is subsiding as the ice (dirty grey colour) melts.

Hoary marmot (*Marmota caligata*) in its typical habitat, a scree slope, in Mount Rainier National Park, Washington, USA.

Mountain goat mother and kid (*Oreamnos americanus*) in Mount Rainier National Park, Washington, USA.

willow. The colonization sequence is complete when coniferous trees have taken over, cutting out much of the sunlight required by the pioneer plants. Grasses and flowering plants sometimes grow on the surface of Alpine glaciers if the debris cover is thick and the ice inactive, while in Alaska, New Zealand and Patagonia even trees grow on debris covering the snouts of stagnant glaciers.

The vegetation provides food for grazing animals. In the Alps the Steinbock or stone ibex and chamois are common, scrambling with ease over precipitous rocky hillsides, while marmots emerging from holes in the ground make a screeching sound to warn their mates at the first signs of danger. Alpine marmots have a 'vocabulary': emitting different sounds depending on whether the dangers are birds of prey above, or predators on the ground.

In the mountains of western North America, Dahl sheep and mountain goats inhabit terrain similar to that of the ibex and chamois, whilst in less rugged terrain moose, elk, bears and other animals frequently forage for food.

Scandinavia is well known for its lemming years, when a population explosion creates the alleged suicidal tendency of the rodents to

Fireweed (*Epilobium fleischeri*) on an alpine meadow near Triftgletscher, Berner Oberland, Switzerland, and alpine forget-me-nots (*Myosotis alpestris*) growing on a lateral moraine of Vadret Pers (Vadret da Morteratsch in the background).

Gemswurz (*Doronicum clusii*).

Alpine tundra with cotton grass, flourishing between roches moutonnées near Oberaargletscher, Switzerland.

Dwarf gentian (*Gentiana nana wulfen*) on the forefield of Glacier de Tsanfleuron, Switzerland.

The west coast of South Island, New Zealand receives several metres of precipitation a year. Glaciers, such as Franz Josef Glacier, are fed by snow at high altitude, and flow down to near sea level, entering the rainforest that drapes the lower slopes.

drown themselves in the sea. Less widely reported is the fact that many of them perish as a result of running up onto glaciers and dying of cold.

Most primitive of all these life forms is red algae, which in summer frequently stain the snow pink. There is even a creature called an ice worm in North America that completes its life cycle on the snow and ice. A small, wingless insect, the glacier flea, also lives exclusively on glaciers in the Alps and North America, living off wind-blown pollen.

The Himalaya are known for the reputed occurrence of a mysterious animal, the Yeti or Abominable Snowman, but except for its tracks, it has proved elusive to western mountaineers. Some Sherpas talk of a hairy, monkey-like animal, but others think the creature is more likely to be a bear or snow leopard. Whatever the truth, it seems that some unidentified animal prowls around the glaciers of the Himalaya, and rarely comes down to lower altitudes.

Thus from the minute to the mythical, glacier country demonstrates the animal kingdom at its most resilient, and those people who venture into such areas may be privileged to see a remarkable variety of life, well-adapted to the hostile environment.

The colonization of terrain after a glacier retreat is shown in an exemplary fashion along a path leading towards the tongue of Vadret da Morteratsch in the Engadin, Grisons, Switzerland. The photographs on the left were taken in 1985, the ones on the right in 2002. The markers indicate when the glacier last reached the respective locations (1970, 1940 and 1900 from top to bottom). Note how much the trees have grown within the 17-year time-span.

12 Benefits of glaciers

To many people, the most obvious benefit of glaciers is their landscape value. This is especially true when visiting **glacierized** areas such as the Alps or Alaska, or **glaciated** areas like the Scottish Highlands and Yosemite National Park in California. However, we may also pose some questions concerning the wider importance of glaciers in terms of the benefits they provide for human civilization. For example, we may well ask: 'How much is a glacier worth?' 'Would it matter if most glaciers in temperate latitudes melted away?' 'How can glaciers, their meltwaters and depositional products be used for our benefit?'. Therefore, in this chapter we explore not only the benefits of today's glaciers, but also those that have long since disappeared.

Mark Twain tells a story in the late nineteenth century of how he travelled to Zermatt, took the mountain railway up to Gornergrat, a well-known viewpoint from which tourists can observe the Matterhorn, and then sat down on nearby Gornergletscher. His plan was, at least as far as the story goes, to use the glacier's motion to return to Zermatt, thus saving the return fare on the railway. Perhaps this was one of the first attempts to cite the value of glaciers, albeit in a rather amusing way.

Irrigation and energy supply

A number of the world's desert regions, such as northwestern China, the Thar Desert of north-western India and Pakistan, the coastal desert of Peru or the wine-growing area of Mendoza in Argentina, all receive waters from adjacent, glacierized mountain ranges. For example, Himalayan meltwater and snowmelt are fed into the Rajasthan canal system, and through hundreds of kilometres of irrigation channels life is brought to the Thar Desert. Some of

One way of letting casual tourists enjoy glaciers is to excavate ice caves or 'grottos'. This tunnel is beneath the Rhonegletscher, Switzerland.

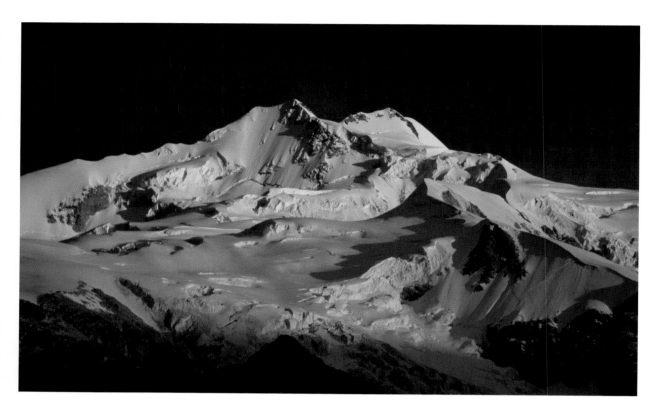

Many people depend on glaciers for their water supply. The world's highest capital city, La Paz in Bolivia at around 3700 metres relies on water from the nearby Cordillera Real, as its immediate surroundings are arid. The state of health of these glaciers in a warming climate is therefore a matter of concern. The mountain illuminated by the setting sun is Huayna Potosi (6088 metres).

this water is runoff from glaciers – perhaps not a large contribution, but one that is most needed in hot, dry years.

Glacial meltwater is important for irrigation in the Alps. Central valleys, such as the Rhône in Switzerland, receive very little precipitation because of the rain-shadow effect of the mountains from humid air masses originating in the north and the south. Thus water shortages limit the productivity of arable land. However, plenty of snow falls on the mountains nourishing many glaciers. An intricate network of irrigation channels has been built and maintained over hundreds of years in order to bring glacial meltwater to agricultural areas. It was necessary for some of the channels to be installed in quite inaccessible places; for example, some made from wood were fixed to vertical rock faces. Meltwater in these channels carries fine sediment which blocks holes in the channels and makes them watertight, but also provides a rich supply of minerals to the pastures and fields that are being irrigated.

A complex series of irrigation ditches, as here on the slopes of Huascarán, Cordillera Blanca, Peru, supports extensive agriculture during the dry season. Most water comes from the nearby glaciers.

Agriculture has never been big business in alpine regions, but hydroelectric power generation is a major sphere of economic activity in several countries, notably in Norway and the Alpine nations. In Austria and Switzerland many power stations were built during the 1950s and 1960s. So much of the available water flows through tunnels and pipelines that the loss of many former mountain torrents and even waterfalls is the sad result of civilization's need for electrical energy. Dams retain the runoff from snow and glaciers during the summer melt-season, when electricity consumption is lower. Then during winter when demand is greatest, half of Switzerland's energy production is generated by water released from these reservoirs. The Massa hydroelectric power

In some areas of Switzerland, the waters are collected from several glaciers and sent via tunnels to one holding reservoir. Here the meltwater from Haute Glacier d'Arolla enters an intake and is then transferred via a tunnel into another valley. In the background is Bas Glacier d'Arolla with a prominent icefall. The main drawback of using glacial meltwater is the sediment in transport. Larger stones are trapped at grids, but the sediment in suspension causes wear on turbine blades. They need to be reshaped or replaced after a certain time in use.

station near Brig, owned by the Swiss Federal Railway System, runs almost entirely on meltwater from the Grosser Aletschgletscher, the largest glacier in the Alps. Water from glaciers is most abundant in warm years, and remains reliable even when there is drought. At such times the glaciers provide the bulk of the energy needed to power the Swiss electrified railway system.

Some glaciers have sometimes been a hindrance to the engineers who tap their water. During the 1970s some large alpine valley glaciers advanced by several hundred metres. Intakes through which meltwater is transferred into the reservoirs became blocked by the advancing ice. However, such problems now belong to the past, because a general retreat on a massive scale has totally shifted the emphasis of concern.

Tourism

A number of large alpine tourist resorts such as Zermatt, Saas Fee and Sankt Moritz in Switzerland, and Chamonix in France make considerable use of the neighbouring glaciers. High firn basins provide extensive areas of relatively crevasse-free surfaces that have been developed for summer skiing. In years when winter snowfall is delayed or scarce, and the traditional skiing areas are starved of snow, the summer ski areas are also put to good use in winter.

Some summer skiing areas were established in the 1970s on a number of small mountain glaciers in several other resorts in the Alps. Unfortunately, these small ice masses have been particularly vulnerable to climatic warming. Thus the glaciers on Titlis near Engelberg, and at Diavolezza near Pontresina have almost totally disappeared, putting an end to summer skiing.

Compared with the much more important winter skiing, the loss of summer skiing opportunities may appear to be of marginal importance. Continued de-glacierization will, however, severely affect the scenic attraction of high alpine ranges such as the Bernina mountains in the Engadin, the Bernese Oberland or the Valais.

Of the glacier attractions in the Alps, the Rhonegletscher is perhaps the best known, but may soon be of little economic value. For many years travellers have broken their journey over the Furka

Fog lends an eerie atmosphere to tourist boat trips on proglacial lake Jökullsaarlon, where icebergs have calved from receding Breiðamerkurjökull, southern Iceland.

The Rhonegletscher ice grotto entrance site, Switzerland. The melting back of the ice margin, typically several metres a year plus the lateral shift due to glacier flow, causes old tunnels to be abandoned, leaving the eyesore of abandoned walkways, but a fascinating experience can still be had when inside.

Pass in order to visit the beautiful blue glacier tongue close to the road. Year after year, the owner of a nearby hotel excavated an ice tunnel as a special attraction for his guests, although because of the lateral shift that occurs as a result of glacier flow, a new tunnel needed to be cut each year. Through this tunnel tourists could enter the inside of the glacier, experiencing the blue light filtering through the ice, the strange sensation of freezing air on a hot summer day, and (last but not least) a bar at the end of the tunnel. However, the days of the Rhonegletscher ice tunnel may be numbered. By 2002 the ice had receded so far back from the valley side that it has become almost inaccessible. Tourists were still able to reach the tunnel by crossing wooden bridges and other contraptions, but it will soon become impracticable to maintain this infrastructure.

Commercial value of glacier ice

Before refrigerators were invented, glacier ice itself was an export commodity for Norway, and allowed glacier-free countries like

Great Britain to keep meat cold. In the Cordillera Blanca of Peru a few people make a living out of collecting glacier ice, grinding it up, mixing it with flavouring, and then selling it at the markets as a local version of ice-cream. Donkeys and mules are used to transport the neatly cut blocks from nearly 5000 metres above sea level down to the valley. To slow down melting under the influence of the tropical sun and the animal's body heat, the ice was packed in thick layers of grass.

In recent times some scientists have considered the feasibility of towing Antarctic icebergs to regions with hot arid climates, such as Australia, north Africa or the Arabian peninsula, in order to provide clean drinking water. At times, Antarctic icebergs drift a considerable distance north under the influence of wind and currents, sometimes even as far as the Cape of Good Hope and Cape Horn, so the final towing distances may be less than at first expected. However, considering that the draught of a typical Antarctic iceberg is several hundred metres, the real problem will be where to berth it. Many icebergs have a draught that exceeds the depth of low-latitude continental shelves. Thus iceberg-towing to desert areas may remain an unrealistic goal for logistical and economic reasons, at least for the next few decades.

A major tourist attraction near the town of Huaraz in Peru is the ice cap of Pastaruri, at a height of over 4500 metres. Visitors are encouraged to take the available horses to minimize the effects of the high altitude.

Products of glacial deposition

There is a clear economic value in the sediments that ice age glaciers have left behind in or near densely populated areas. During the last ice age, 30 per cent of the world's land area was ice-covered, including much of northern and central Europe, central Asia, North America and various countries in the southern hemisphere. Deposits released directly from glacier ice are poorly sorted mixtures of mud, sand and gravel, and their value is in providing mineral-rich soils that are ideal for agriculture. In addition, in countries like Canada and Finland they allow the tracing of valuable minerals back to their source. We have also seen how glaciers have produced large volumes of water that typically sort the sediment. Such sediment is referred to as glaciofluvial, and provides abundant sand and gravel resources – ideal for the construction industry. Large sand and gravel quarries are scattered around glaciated lowland areas, such as the Swiss Plateau, northern Germany and Poland, the Great Lakes region of North America, and the Midlands and East Anglia in England. The resources from these quarries are so easy to extract that they can be carried hundreds of kilometres before extraction

Glacier snouts can be very dangerous, especially if they are advancing. This sign at Franz Josef Glacier, South Island, New Zealand warns tourists of the dangers of the collapse of the glacier portal, and the whole glacier frontal area is cordoned off.

With many of the world's mountain landscapes being fashioned by glaciers, it is instructive to take students to areas with modern glaciers so that they can appreciate the processes forming the landscape. Here, a Swiss high-school student is surveying at Vadret da Morteratsch near Pontresina, southeast Switzerland. Using modern laser ranging equipment it is possible, even for students, to measure ice flow over only a day or two.

and transport costs exceed the value of the material. Elsewhere, in highland areas small sand and gravel quarries serve local communities or even just individual landowners.

The value of a sand and gravel pit does not end with the removal of the commodity. Abandoned pits are much sought after as landfill sites. However, much care is needed to prevent fluids from the pit seeping from the glaciofluvial sediment into the local water table, and therefore knowledge of the sediment is important.

Water resources

Sand and gravel, being permeable, commonly serve as an underground reservoir or **aquifer**. In some areas of England, such as around Liverpool, the water in sand and gravel is believed to have accumulated during the last ice age some 20000 years ago. At this time large rivers were generated at the edge of the receding ice sheet and provided a good source of water. This and other sand and gravel bodies are topped up with rainwater, and have the beneficial effect of filtering the water.

On a larger scale, lakes occupying glacial troughs serve as vital reservoirs in many countries. Glacial lakes such as the Finger Lakes in New York State or indeed the Great Lakes are invaluable reservoirs, containing huge volumes of fresh water. Water from Bodensee, a large lake gouged out by the Pleistocene Rhine Glacier, is pumped through hundreds of kilometres of pipeline in order to provide drinking water to the southern German provinces of Baden-Würthemberg and Bayern. Millions of people depend on this fresh water resource. Those countries sharing the Bodensee catchment exercise strict control of water quality. In Britain lakes occupying glacial troughs are often dammed to increase their capacity. For example, lakes in Wales serve English cities such as Birmingham and Liverpool, while the Lake District has several lakes that serve Manchester. Other, usually dammed, glacial valley lakes are used for the generation of hydroelectricity, particularly in the Scottish Highlands. Whether the water is for drinking, industry or power generation, the construction of supply facilities is invariably controversial, and planning authorities have to balance such needs against the damage to the scenic qualities of the area concerned.

Scenic value of glaciers and their landscapes

Adventure tourists have sought after glaciers ever since the pioneering mountaineers visited the Alps in the early nineteenth century. Nowadays, many glaciers are reached by cable cars or funicular railways, giving almost anyone access to magnificent glacial landscapes. For instance a cable car takes visitors across the icefields high up on Monte Blanc from France into Italy. The mountain railway from Zermatt to Gornergrat in Switzerland gives people splendid vistas of the Gornergletscher and famous peaks such as Monte Rosa and the Matterhorn. Cruise-ship passengers can now travel in comfort to see the calving glaciers of Alaska, Svalbard, Patagonia and even Antarctica.

Landscapes shaped by Pleistocene glaciers ten thousand or more years ago are also visually attractive. Time has allowed vegetation to cover the untidy heaps of glacial debris. Lakes dammed by terminal moraines such as Lago di Garda and Lago di Como at the foot of the Italian Alps, or the many beautiful lakes in Austria's Salzkammergut are the product of those times that Alpine glaciers reached out into the lowlands. The ice-carved basins of the Lake District in England, the Scottish Highlands and North Wales provide us with Britain's most attractive scenery. The mountains of western North America, such as Glacier National Park and Yosemite, provide excellent recreational opportunities. The once glacier-occupied fjords of Norway, British Columbia, South Island of New Zealand, Chile and western Scotland provide tourists with beautiful vistas in near-natural landscapes.

Glaciers as an 'educational resource'

In concluding this chapter we draw attention to what could be described as the 'educational value' of glaciers. Both authors have led many field trips for students, during which the participants have studied glaciers and glacial landforms. For many years, Jürg Alean has taken students of the Kantonsschule Zürcher Unterland at Bülach, a Swiss secondary school, to Vadret da Morteratsch, a valley glacier in the Canton of Grisons. During an intensive week of fieldwork, they become acquainted with glacial phenomena both on, and in front of, the glacier snout. Groups of students have documented year after year the position of the glacier terminus, as well as changes in the vegetation in front of the receding glacier tongue.

Apart from being spectacular and adventurous, this kind of fieldwork offers an invaluable insight into the dynamics of glaciers and the processes of erosion and deposition. Furthermore, the work provides the students with an appreciation of the grave consequences of climatic change. Whereas most geological processes are much too slow to be seen 'live', and thus leave much to the imagination, developments on and around a glacier take place at a more 'reasonable' speed. Students in fact can measure movement of glacier ice in no more than two or three days using laser distance measuring equipment. Again, within no more than one year, the glacier tongue has usually receded sufficiently for simple positional measurements, using a hand-held GPS receiver, to clearly show the changes that have occurred within that year.

Terrain in front of a glacier also provides a natural laboratory for life scientists. Again the situation at Vadret da Morteratsch is ideal: snout positions are clearly marked by prominent signposts that date back to 1900. Many a student has found it hard to believe just how far the ice has receded, whether it be over the full century, the last decade or simply the past year. With recession comes colonization of the bare ground, first by alpine plants and then by shrubs and coniferous forest. By investigating how new plant species are introduced and how quickly they grow, students can acquire an understanding of the ecology of the alpine environment.

For students taught by Michael Hambrey at the University of Cambridge, Liverpool John Moores University and the University of Wales in Aberystwyth, there have been fewer opportunities to study 'living' glaciers, but the UK does have a rich heritage of glacial phenomena, dating from the last ice age. Excursions to mountains such as Cadair Idris in mid-Wales introduce students to some of the country's classic glacial erosional landscapes, whilst sediment exposures along the Welsh coastline or in East Anglia give them practice in recording the properties of glacial sediments that is so important in careers involving sand and gravel exploitation, waste disposal, water resource evaluation or environmental issues. Some students may undertake a glacier-related project for their final year undergraduate dissertation, such as glacier hydrological studies near Arolla in the Swiss Alps, or glacial depositional process investigations in Svalbard.

A small proportion of these students will do well enough in their undergraduate degree to undertake postgraduate studies in glaciology or glacial geology, for which there is hardly any geographical limit, as long as there is a member of staff with the experience and regional knowledge to advise them.

It was the fascination with the beauty of ice-dammed lakes,

One of the major benefits of glaciers in developed countries is their ability to provide a reliable water supply for hydro-electric power generation. Here a dam at Griesgletscher, Valais, Switzerland has been constructed to receive meltwater in summer for use in power generation in winter. During the extremely hot summer 2003 so much meltwater was available that power generation continued throughout the summer, when there was a shortage of electrical energy in central and southern Europe.

glacier tables, or towering glacial peaks, but also with chilly depths of a wide-open crevasse, treacherous moulins or thundering ice avalanches that brought the authors into the field of glaciology. They both continue to enjoy sharing these experiences with students, and it is very satisfying when they choose a career in the Earth sciences, and add their own contribution to the understanding of glaciers, and their resources and response to environmental change.

13 Glacier hazards

Over the centuries glaciers have been responsible for many disasters. In the worst affected area, the Cordillera Blanca in Peru, over 32 000 people lost their lives in glacier-related disasters during the twentieth century alone. We have already described the perilous combination of glaciers and volcanoes (Chapter 9), but there are other ways in which glaciers can cause havoc, notably by ice avalanches and bursts from glacial lakes. Such phenomena are quite common, but only constitute a hazard when human lives or property are at risk. As populations increase in glacierized mountain regions such as the Andes and the Himalaya, more and more people are becoming exposed to glacier hazards. Furthermore, glacier recession is leading to the development of an increasing number of dangerous glaciers, and mitigation measures are needed to prevent future disasters. In this chapter we focus on the main types of glacier hazard, giving examples of several of the disasters that have befallen mountain communities, explaining how mitigation measures are undertaken in order to prevent future catastrophes.

Although glacier disasters are not usually as dramatic as, for example, a major earthquake or volcanic eruption, the long-term consequences may be severe, especially for countries with fragile economies. Not only may there be loss of human life and property, but economic consequences may include disruption of transport infrastructure, damage or destruction to hydro-electric power schemes and irrigation, and lost production – all of which can run into hundreds of millions of dollars. Few governments in developing countries are able to afford preventative measures, but United Nations organizations and the aid agencies of wealthy countries are taking an increasing role in encouraging glacier-hazard evaluation, as well as supporting mitigation measures, albeit on an inadequate scale. However, we need to recognize the type of hazard in the first

A curious block of ice has fallen off a hanging glacier on the south side of the Mönch (4099 metres), Berner Oberland, Switzerland. Despite its moderate mass of about 55 tonnes, the compact snow surface has allowed it to travel an exceptionally long distance (half a kilometre) down the Grosser Aletschgletscher. Falling ice is one of the dangers that climbers face in glacierized mountain areas.

place – a challenging problem when the hazard may be in inaccessible high-mountain regions.

Ice avalanches

Characteristics of ice avalanches

Ice avalanches represent falls of ice (often with rock) on a time-scale of minutes. Scientists have recognized three main trigger mechanisms: peeling off of a block of ice from an unstable cliff, failure of a slab of ice from a steep glacier induced by sliding at the bed, and more deep-seated failure involving bedrock as well as the overlying ice.

A major problem in predicting ice avalanches is that, despite their spectacular impact, they are relatively rare – much rarer than snow avalanches. Nevertheless, dozens of large ice avalanches in the Alps and North America have been detected on thousands of aerial photographs taken especially for this purpose, and the data resulting from the mapping of these avalanches have then been used to predict, at least approximately, the distances they are likely to travel, that is

Ice avalanche from the snout of Festigletscher, Canton Valais, Switzerland in 1981. An estimated 2000 cubic metres of ice were involved. Considerably larger events of this nature have been known to threaten the village of Randa in the valley below.

A major ice avalanche from a hanging glacier on the south flank of Mönch (4099 metres), Berner Oberland, Switzerland. More than 300 000 cubic metres of ice fell, some of it traversing the popular route from Jungfraujoch to the Möchsjochhütte. Nobody was hurt, but these tourists are unknowingly exposing themselves to further potential avalanches as they walk across the grooves left by the sliding ice masses.

their **run-out distances**. However, it is difficult to predict the size of an avalanche and the distance it will travel. This is partly because the roughness of the terrain has a complicated and variable braking effect on the falling ice mass. For example, the avalanche path may be very rough in summer and slow the avalanche down more effectively than in winter, when the path is covered with snow. Often, the ice does not break off the glacier in one piece, and smaller chunks may fall off weeks before the major event.

Glaciologists have attempted to find ways of predicting the ice volume to be released in avalanches, as well as the timing of the event. It is now known that the ice in the unstable part of a glacier accelerates drastically prior to break-off, usually creating fresh crevasses. However, in practice it is very difficult to monitor these developments, since they occur most frequently at high altitudes.

Ice avalanches in the Alps

On 31 August 1597 a mass of ice detached itself from the Balmengletscher, fell onto the village of Eggen and buried it, together with 81 people, all cattle and all other possessions. Fortunately, many inhabitants at that time were still working on the alpine pastures; otherwise the disaster would have been even greater. Subsequently, the pile of ice took seven years to melt away. This is the earliest historical record of a catastrophe caused by an ice avalanche. The village, then situated near the Simplon Pass in the canton of Valais in Switzerland, no longer exists, and neither does the glacier, having disappeared last century as a result of climatic warming.

However, the threat of ice avalanches from glaciers persists, both in the Alps and in other densely populated, glacierized mountain ranges. The Valais, in particular, has a sad history of ice-avalanche catastrophes. One of the worst occurred on 30 August 1965 during the construction of a dam for a hydroelectric power plant at Mattmark in the Saas valley, when a slab of ice, with a volume of about a million cubic metres, broke away from the tongue of the Allalingletscher. Within seconds the avalanche had swept over part of the construction site burying the workers' camp and killing 88 people.

In some instances a glacier may provide no prior warning of an impending catastrophic ice avalanche. In 1895 almost an entire glacier slid off the steep northwestern face of Altels, a prominent peak south of Kandersteg in the Bernese Oberland. More than four million cubic metres of ice fell down, and six shepherds and many cattle on an alpine meadow were killed. Calculations have shown that the ice must have reached a speed of 400 kilometres an hour as it swept across the route to the Gemmi Pass, some of it ending up nearly 400 metres above the valley floor on the opposite side.

Ice avalanches in the Cordillera Blanca, Peru

Far more devastating ice avalanches have occurred in the Peruvian Andes, particularly those involving Nevado Huascarán (6768 metres), Peru's highest mountain. At the foot of this heavily glacierized moun-

The glacierized summit of Nevado Huascarán Norte (6746 metres) at sunset. It was from this face of the mountain that the world's most devastating ice avalanche took place on 31 May 1970. Triggered by a powerful earthquake, part of the summit icefield collapsed along with rock from the exposed face below the summit. The avalanche swept down the mountain, obliterating the town of Yungay and its 18000 inhabitants.

tain lies the beautiful valley of the Rio Santa. In 1962 a colossal ice and rock mass became detached from the precipitous north summit: 4000 people were killed in the ice avalanche.

Only eight years later, on 31 May 1970, an even worse disaster struck, when a powerful earthquake triggered a second, much larger deep-seated avalanche. Heavily fractured rock from the west face of the mountain broke off, carrying with it part of the glacier that it supported. Approximately 50 million cubic metres of ice, rock, moraine material and water shot downwards, covering the 16 kilometres to the valley bottom in no more than three minutes. Because of its larger volume, this avalanche spread much further laterally than the earlier one, and part of it leapt over a ridge and buried the town of Yungay. About 18000 people were killed instantly adding to the tragic death toll of the earthquake. In total 70000 people lost their

lives. The following account by a well-known Peruvian glaciologist, Alcides Ames, who was evaluating remediation measures associated with the moraine-dammed lake of Laguna Safuna high in the Cordillera Blanca at the time, indicates just how harrowing this event was for the local people.

On 26 May 1970, I was sent with three others by the Glaciology Division of Corporación Peruana del Santa to the Laguna Safuna area to carry out several complementary surveying tasks after the tunnelling works had been finished. The workers had left three weeks earlier, and only two men remained as camp watchers. The following Sunday, 31 May at 3.20 p.m., the most devastating and powerful earthquake of recent decades happened, reaching 7.8 on the Richter scale and lasting 48 seconds. At first we all thought that the shake was local, bearing in mind that in 1945 an earthquake happened in the same area. Soon after the shock of the trembling ground, and seeing many big avalanches from the high walls of the Pucahirca mountains, we all decided to inspect the recently built tunnel, and found that it was severely tilted from its former horizontal alignment. The reinforced concrete was broken in several places, rendering the tunnel completely useless for the purpose for which it had been built.

Fortunately, we had brought with us a small radio receiver. In the evening we tried to listen to broadcasts from Lima to get news about the earthquake. At 11 p.m. we heard that the Santa valley and several cities on the coast had been severely damaged. There were thousands of victims, the roads were broken or blocked by landslides. In Huaraz, my home town, 90% of the houses had collapsed. The town of Yungay had disappeared, buried by tons of mud, the consequence of an ice and rock avalanche from the trembling north peak of Huascarán that jumped the hill that apparently protected the town. Hearing this terrible news, we decided to go back home to see what happened to our families. Since landslides now blocked all the roads, we agreed to go over a high mountain pass that we knew very well. Early on the Monday morning we left for the 5300-m-high col between the peaks of Alpamayo and Pucahirca. We had only mountaineering boots and sleeping bags, and no ice axes, crampons or ropes. However, we had already met a group of eight New Zealand climbers, and after visiting their camp, they agreed to lend us their climbing equipment and to accompany us, so that they could carry their things back from the other side of the pass.

Opposite. The full length of the route of the disastrous Huascarán avalanche is illustrated in this photograph taken from the opposite side of the valley in 2002. The avalanche was largely constrained by the steep valley below the avalanche site at first, but mixing with moraine debris and water it became a fast-flowing debris-flow. The main tongue, emerging from the confines of the steep valley, spread out as a huge fan, temporarily blocking the river at its foot. A secondary tongue spilled over a ridge to the left of the steep valley, wiping out Yungay, now represented by the dark brown patch and a memorial park (far left middle of picture). Yungay has been rebuilt in a safer location on the hillside off the picture to the left.

On Monday 1 June we left the New Zealanders' camp at 10 a.m., climbed up to the pass and descended by the Arhueycocha valley into the main Santa Cruz valley. On the way down we saw big landslides still falling from the steep slopes of the canyon-like valley to the west. Despite being unsure about whether we had chosen the right way to go, we continued our journey, arriving at a small hut beside the Hatuncocha lake in the dark at 8 p.m. We had managed to avoid rock slides on the left side of the valley, although one of us was almost hit by a flying piece of rock.

After spending the night at Hatuncocha, we departed early in the morning, very worried about the landslides further down the valley. Anyone who has walked through the Santa Cruz valley can appreciate that the entrance is a very narrow gorge, so when we were approaching it we saw clouds of dust from landslides rising up. On arriving at the first landslide we waited for rock to stop falling before running quickly over unstable boulders. This procedure was repeated three times before we reached the village of Cashapampa, completely covered in dust, very dirty and sweaty. The village people were very surprised that we had come out safely, and they were very friendly and generous in giving us some food to eat. We continued our way to Caraz, arriving there at 6 p.m., where we found colleagues from our company who offered us overnight accommodation and food.

On Wednesday 3 June we spent the morning trying unsuccessfully to persuade one of the company's engineers to take us by helicopter to Huaraz. We were told that landing in Huaraz was quite impossible due to dust in the air from the collapse of mud brick-built houses. So at noon, three of us continued walking to Huaraz, our other colleague staying at Caraz, being exhausted. We arrived at Yungay late afternoon, seeing how the avalanche had buried the town. We shed some tears, thinking how so many people had lost their lives in just a few minutes. It was the saddest and most shocking experience that any of us had ever had. It was now too late to use the temporary bridge across the Shachsha River, so we spent the night in a safe place between the buried Yungay and Ranrahirca, where, as at Cashapampa, the few survivors were camping in fields and generously gave us some food. That evening, at 11 p.m. another strong shake occurred, producing minor avalanches that increased the discharge of the river.

Early in the morning of Thursday 4 June, we were told by the bridge builders to carry some pieces of wood to the site. Walking in bare feet, we sank up to our knees in the mud. Crossing the river, we were fortunate in finding that the road between Ranrahirca and Carhuaz was not blocked, so we took a pick-up truck to shorten the trip to Carhuaz. From here we continued walking, arriving at Huaraz at 5 p.m. after more than 100 km of walking, safe but exhausted. Near the town we met one of my wife's friends, who informed me that she and my children were safe. I found them at my mother's house, which fortunately had not collapsed, but our house in the town centre had collapsed completely. Later my companions told me that they had also lost their homes, but that their families were safe. I noticed that in the narrow streets of the town thousands of people had perished and many were still buried beneath the debris. I experienced the same feelings as at Yungay, thinking about the many people who had died, but personally I was happy to find my family safe.

Next day I went back to work, where my boss told me that I had four days to manage my private affairs. After this I returned to the office because there was so much work to do over the next few months. About 8 days after my departure from Laguna Safuna, I resumed work, and was sent to Llanganuco lake to survey it. A minor ice and rock avalanche had fallen from the same north peak of Huascarán and dammed the upper lake, raising its level by 8 m. This avalanche killed the fourteen members of a Czechoslovakian expedition camped near the lake as well as an unknown number of Peruvian tourists.

Alcides Ames: Personal account of the 1970 Peru earthquake

From this account, it is clear that settlements in the Cordillera Blanca are vulnerable to ice avalanches, even if other forces such as earthquakes trigger them. The town of Yungay has now been rebuilt in a safer position, and the buried town site is now a moving memorial to that tragic event. Unfortunately, because of lax planning regulations, new homes have been built on top of the avalanche debris, and are thus vulnerable to any future avalanche from the summit of Huascarán.

A recent ice-avalanche disaster in the Caucasus Mountains, Russia

Settlements exist close to high, steep and glacierized peaks in the Caucasus Mountains. One of the strangest glacier catastrophes occurred on 20 September 2002. The sequence of tragic events started with the detachment of a nearly 1.5-kilometre-wide mass of rock and ice from the perennially frozen north face of the peak Dzimaraikhokh in the Kasbek mountain massif. From an altitude of 3600–4200 metres, several million cubic metres of ice and rock crashed down onto the gently graded tongue of Kolka Glacier. Although this glacier had surged in the past, such as in 1969, it was not in a surging state in 2002. The entire lower part of Kolka Glacier was sheared off by the impact, thus contributing tens of millions of cubic metres of ice and moraine material to the avalanche. Never before have glaciologists recorded the detachment of such a flat glacier tongue.

The avalanche then crossed the tongue of Maili Glacier, picking up more debris. With a speed of around 100 kilometres per hour it then travelled a further 18 kilometres northwards through a deep valley towards the village of Karmadon, where it killed dozens of residents. As the valley narrows north of Karmadon, most of the avalanche was arrested and roughly 80 million cubic metres of ice and debris were deposited. However, mud and water were squeezed out of the avalanche and travelled another 15 kilometres through a narrow gorge below Karmadon, killing more people along the way. In total, the avalanche and mudflow claimed more than 120 victims.

Researchers at the University of Zürich, in collaboration with colleagues of the United States Geological Survey, immediately investigated the devastating effects of the event, using satellite imagery. By quickly reprogramming NASA's TERRA satellite and its ASTER instrument (Advanced Spaceborne Thermal Emission Radiometer), images with a ground resolution of 15 metres were obtained on 27 and 29 September 2002. As the images were taken from different angles, stereoscopic models could be derived, yielding a unique basis for investigating the dynamics of the event.

Images obtained subsequently showed the development of huge lakes, both on and to the side of the avalanche deposit. These now had to be monitored carefully as millions of cubic metres of water posed a high risk of further flooding down-valley.

At present, it is uncertain whether climatic warming had anything to do with triggering the ice- and rock-fall that initiated the avalanche and subsequent mudflow. It is clear, however, that at least parts of the mountain face from which the mass became detached were in a permanently frozen condition. Ice embedded in rock fissures is known from other high mountain regions to have a stabilizing effect. Melting of such frozen ground, or 'permafrost', in steep rock slopes may thus increase the chances of major rock-falls.

Prediction of ice avalanches

Prevention of ice avalanches is normally impracticable, in contrast to the generally highly successful measures that are implemented for snow avalanches. The timing of an event such as the Huascarán disaster, described above, could not have been predicted. We can observe the ice cap on the summit today, noting the growth of crevasses and the state of the rock beneath the ice, and say that another avalanche is likely. However, it is not possible to estimate when the next avalanche might occur, nor the volume of ice likely to be involved, nor the run-out distance, without detailed monitoring at considerable expense.

In contrast, in a relatively wealthy country like Switzerland, prediction of ice avalanches has become a rigorous science. Hazardous glaciers are monitored using remote-sensing techniques, repeat photogrammetric mapping, ground surveying and mathematical modelling of ice dynamics. Through such investigations, the attributes of dangerous glaciers that are determined include: potential break-off volume, geometrical changes of the glacier at its front, and ice-recession or advance over ground that is likely to induce collapse.

One of the few in-depth studies of an unstable glacier that led to accurate prediction of an ice avalanche was undertaken on the Weisshorn (4505 metres) in Switzerland, a mountain with a history of releasing catastrophic ice avalanches onto the village of Randa below. Following the recognition of opening crevasses on the glacier in 1970, a monitoring programme was initiated by glaciologists at

Satellite image of the town of Huaraz (H) and its relationship with the nearby Cordillera Blanca, Peru. The town experienced several natural disasters in the twentieth century including an outburst flood from glacial lake Laguna Cohup (P) on 13 December 1947 when 7000 people died. The authorities are keen to mitigate glacial hazards before a new disaster strikes. One example is the lowering of the moraine-dammed lake Llaca (L) in the adjacent valley. (Image from ASTER sensor aboard NASA's TERRA satellite; image taken on 5 November 2001; source: http://earthobservatory.nasa.gov/Natural Hazards/natural_hazards_v2.php3?img_id=10128).

the Swiss Federal Institute of Technology, with the installation of automatic cameras on nearby vantage points. Photogrammetric maps were prepared as the geometry of the glacier changed and crevasses grew, and ice velocities were determined. The snout area of the glacier showed a slow increase in velocity at first, but it accelerated rapidly from late 1972 onwards. It was clear by August 1973 that the steepening velocity curve was indicating imminent collapse, and an ice avalanche was duly predicted accurately for 19 August. Appropriate warnings were issued to the villagers in Randa 3000 metres below. In the event, the avalanche took place in two stages, and ice failed to reach the valley floor, but the exercise proved the value of close monitoring of dangerous glaciers on steep slopes. Unfortunately, for those countries where major avalanches are a problem, the resources and expertise do not exist to monitor dangerous glaciers in the detail that is required to enable predictions to be made.

Outburst floods from glaciers

Characteristics of outburst floods

A second major category of glacier hazard involves a broad group of processes culminating in floods, which take place on a time-scale of hours to a few days. Glacier outburst floods refer to the rapid discharge of water from within a glacier or from an ice-dammed lake. Sometimes the Icelandic term **jökulhlaup** is used to describe such floods, although strictly the term applies to the type of outburst flood that accompanies subglacial volcanic eruptions. Glacial outburst floods are characterized by a huge increase in discharge within minutes or even hours, followed by a gradual tailing off. The flood wave may be several metres high and, because of their unpredictability, such floods are extremely dangerous. In the Alps or the Western Cordillera of North America most outburst floods occur in summertime; the lakes gradually fill up as the melt-season progresses until there is sufficient head of water to lift the glacier from its bed, allowing the water to escape suddenly. Such lakes generally remain dry with the onset of efficient drainage in late summer, but channels are subsequently sealed in winter because of ice deformation.

Ground view of lake Llaca illustrating the high moraine dam, behind which is a lake with a calving glacier. The spillway was constructed to lower the lake to a safe level.

Given that the temperature of the ice can affect the development of the drainage in a glacier, it is not surprising that ice-dammed lakes are most common at the margins of cold or polythermal glaciers where the internal drainage system is less effective. Ice-dammed lakes in the Polar Regions are larger and more common than those in temperate regions, but as they are far removed from centres of population, they do not create the same havoc when they burst.

A different type of flood results from the failure of a moraine-dammed lake in front of a receding glacier, particularly in the Andes and Himalaya. The terms **debâcle** (French), **aluvión** (Spanish) or **glacial lake outburst flood** (GLOF) are used to describe these events. The moraines are usually huge ridges of debris built out from the valley sides, and sweeping round in an arc across the valley, defining the latest maximum position reached by the glacier. Commonly, waters which breach the moraine dam contain huge amounts of debris, including boulders several metres across, derived from the moraine itself, with the flood taking on the appearance of a fast-flowing mud-flow. Various processes can lead to the failure of a moraine dam. Firstly, the dam may have a core of dead glacier ice that melts slowly beneath its carapace of debris. Melting slowly lowers the effective height of the dam, allowing the water to drain into the core and weakening it. Secondly, the dam may be overtopped and weakened by large waves, triggered by calving from the receding glacier front, or by ice or rock avalanches falling into the lake. Thirdly, the dam may settle with time, particularly in earthquake-prone areas, lowering the freeboard of the moraine to a level where lake water spills over the dam and begins to erode it, leading to eventual failure. Fourthly, as a glacier recedes, a lake may form on top of it, eventually allowing the submerged ice to become buoyant. When this happens, subglacial drainage may occur instantaneously, eroding the moraine dam and releasing the lake waters. Lastly, the very engineering works that are designed to prevent it can induce failure, particularly when ice cliffs and moraines are destabilized, resulting in collapse and therefore generating destructive displacement waves.

Glacial lake outburst floods in the Cordillera Blanca, Peru

The Cordillera Blanca was the site of the worst ever-recorded glacial lake outburst disaster. It took place in the same mountain valley in Peru that experienced the terrible ice avalanches referred to above. On 3 December 1941, the busy market town of Huaraz was partially destroyed by a flood and at least 6000 people perished. This outburst was from a glacial lake dammed by an unstable moraine. Many such lakes occur in the Cordillera Blanca, since numerous unstable terminal moraines have been left behind as the glaciers have receded since the Little Ice Age of about 200 years ago. Many lakes occupy natural basins which fill with rain and meltwater. Because they contain loose debris, the moraines are easily eroded, particularly during thunderstorms or at times of strong meltwater discharge, or when ice or rock falls into the lake. As the outlet channel deepens, increasing volumes of water rush through it, thereby eroding the

This huge pile of boulders and logs forms a partial blockage of the river below Laguna Soler in a remote valley below the North Patagonian Icefield. We interpret this as the product of a glacial lake outburst flood, as the lake level appears to have been dramatically lowered by a breach in the nearby moraine dam. Boulders scattered downstream are too large to have been moved by anything but a catastrophic lake outburst flood. The remoteness of the site and the ability of a large lake downstream to absorb the flood meant that there was little impact on human civilization in the area.

Figure 13.1 Examples of hazardous moraine-dammed lakes in the Cordillera Blanca, Peru. The debris-covered glacier Hatunraju, which dams Laguna Parón, has been partly remediated by a tunnel in the bedrock. Up-valley to the right the moraine damming Artesoncocha was breached in 1951 following an ice avalanche into the lake. The un-named glacier to the south of Artesoncocha and Hatunraju have the potential to develop lakes inside their moraines as they recede. (Adapted from Reynolds, J. M. *Geohazards, Natural and Man-made.* McCall, G. J. H., Laming, D. J. C. & Scott, S. C., (eds) London: Chapman & Hall, 1992.)

moraine at an ever-increasing rate, until the lake literally bursts out of the confining moraine.

Peruvians, sometimes with outside aid, have made strenuous efforts to remove some of the danger spots. They have transported equipment up to altitudes reaching above 4000 metres in order to construct artificial channels or tunnels.

Perhaps the most impressive example of lake remediation has been the lowering of Laguna Parón. Originally covering an area of 1.6 square kilometres, with a volume of around 75 million cubic metres, this lake is dammed by a debris-covered glacier named Hatunraju, bounded by moraines 250 metres high, as the accompanying map shows. Before remediation, water from the lake drained through the

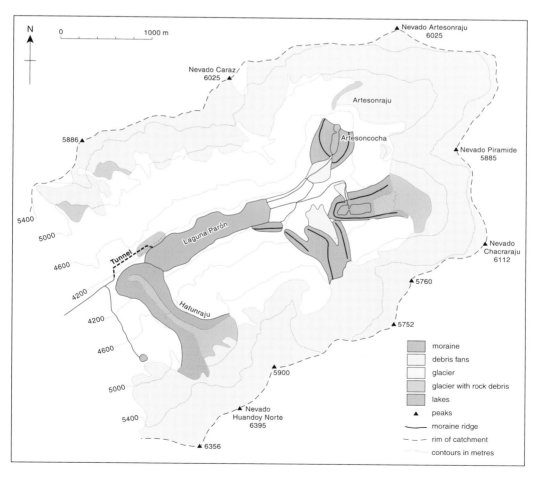

moraine, forming small springs on the down-valley side. Until 1952 the lake level was within two metres of the crest of the moraine. The 'freeboard' was almost over-topped in July 1951, when four to five million cubic metres of ice fell from the sheer ice front of Artesonraju into another lake, named Artesoncocha, at the head of the valley. This ice avalanche triggered the release of over one million cubic metres of water, breaching the moraine dam of Artesoncocha, and lowering it by seven metres. At the same time the level of Laguna Parón was raised by a metre. Continued erosion then led to total failure of the Artesoncocha moraine three months later, releasing a further 3.5 million cubic metres of water. In the second event Laguna Parón was raised by two metres. Fortunately, and unusually, the lower moraine belonging to Hatunraju, close as it was to being breached, held back these large volumes of water, and actually prevented flooding downstream.

Following these events of 1952, the moraine of Hatunraju was raised artificially with loose debris as a temporary measure. A more permanent solution was the excavation of a tunnel through granitic bedrock on the northern side of the lake. This achieved a lake lowering of 20 metres. However, there remains a major uncertainty; there is no way of knowing how stable the moraine of Hatunraju is. Should this moraine fail, an aluvión of around 50 million cubic metres could sweep downstream and severely damage the town of Caraz, 16 kilometres away.

Several other potentially hazardous moraine-dammed lakes have been remediated in the Cordillera Blanca. One that was remediated in 1988 in the knick of time was a lake at Hualcán, where an ice-cored moraine was rapidly decaying, and water was seeping from the lake behind the moraine through the loose debris. The solution in this case was to lower the water level 12 metres by siphoning. After delays resulting from a lack of funds, siphons were eventually installed and a drainage channel dug, so that by June 1990 the lake level had dropped five metres. Funds then ran out, but whether the lake is now safe remains to be seen. However, the short-term threat to the town of Carhuaz was averted. Another example of remediation

The Dig Tsho event triggered much new research on glacial lakes in the Himalaya, especially by Japanese and British scientists in association with Nepali Government organizations. Other moraine-dammed lakes, such as Imja Tsho in the Khumbu Himal, were identified as being potentially hazardous. The failure of this lake, close to popular trekking routes in the Mount Everest region, would have severe implications for the local economy.

We now have a reasonable idea how moraine-dammed lakes develop in the Himalaya. Down-wasting of the debris-covered glacier tongue is succeeded by coalescence of supraglacial ponds, followed by rapid lake enlargement, the development of a calving ice cliff, and decay of the ice core in the moraine. All this can take place in less than a decade. The scale of the problem is not well known, although there are estimates of 20 potentially dangerous lakes in Nepal and a further 24 in Bhutan, to name just two of the Himalayan countries. Reliable inventories based on recent aerial photography or, if unavailable, satellite imagery are needed. In addition, routine monitoring of lakes in their early stages of development is necessary. Remediation is best undertaken before a lake is fully developed, but rarely are funds available until the problem has become acute.

Currently, the most spectacular example of a dangerous moraine-dammed lake is Tsho Rolpa at the head of the Rolwaling Valley. Tsho Rolpa lies at an altitude of 4450 metres and is fed by calving Trakarding Glacier. By 2002 the lake had grown to 3.5 kilometres in length, 0.5 kilometres in width and was at least 135 metres deep. The moraine dam is some 150 metres high, and has a core of decaying glacier ice. The lake has an estimated volume of 110 million cubic metres, of which a third could be released if the moraine dam failed. Several villages and a new hydro-electric scheme are at risk from a possible flood.

Following investigations by a British company specializing in glacial hazards, in co-operation with the Nepali Government, it became clear by the late 1990s that failure was imminent, and remediation measures were needed urgently. Emergency measures were initiated with the installation of an early-warning system, using sensors that would detect a flood-wave above a certain height, and trigger sirens in the villages downstream. Funding from the Dutch Government was acquired for emergency remediation measures, initially involving siphoning, and then constructing an artificial spillway four metres below lake level. Work began in April 1999, and was completed in 2002. However, this is a temporary solution only as 15–20 metres of lake lowering are required in order to render the lake safe. Even the completion of these preliminary measures was no mean feat, as the equipment for such a project at nearly 5000 metres above sea level had to be manhandled by porters to the site, as helicopters cannot be safely utilized at this altitude.

These examples from Peru and the Himalaya highlight the issue facing developing countries that are prone to natural disasters.

In 2002, Ghiacciao del Belvedere near Macugnaga in the Italian Alps advanced in a surge-like fashion. The Little Ice Age moraine was being breached, trees were being felled, and falling boulders began to threaten tourist infrastructure.

A potential hazard associated with the unusual behaviour of Ghiacciao del Belvedere was the rapid growth of a supraglacial lake, the bursting of which was felt to be likely. Pumps were installed to lower the lake to safer levels.

Firstly, there is a basic lack of knowledge about how high-mountain glaciers are responding to climatic change. Many glaciers are inaccessible and the potential hazard is often not even recognized. Interpretation of aerial photographs by specialists is sorely needed, but even where they exist, they may be decades old, as in Peru, or classified for military reasons. Nowadays, the availability of satellite imagery, such as ASTER, has the potential for evaluating glacier hazards on a regional basis. Although the resolution of ground features using ASTER is much less than for aerial photography, it does have the advantage of being up-to-date and readily available. However, trained personnel are still required, and it is important to validate remote-sensing interpretations by means of fieldwork.

Secondly, there is the issue of resources. It is clear that the developing countries themselves cannot afford to undertake evaluation of hazards, let alone remediate them when they are identified. Hence, funds for this work must come from the richer countries, but although they may be keen to assist following a disaster, they are less enthusiastic about funding remediation projects to prevent those disasters.

The Italian Alps

A particularly interesting and totally unexpected development involving 'surge-like' behaviour of a glacier, and associated lake formation, occurred recently in the Italian Alps. It involves Ghiacciao del Belvedere, a debris-covered glacier fed by very steep tributaries descending the precipitous east face of Monte Rosa. The glacier has a long history of outburst floods that have had various causes. In some years, heavy rainfall has led to the accumulation of water pockets that eventually burst from the ice; at other times melt- and rain-water have breached and eroded the high lateral moraines; and finally three outburst floods from moraine-dammed Lago delle Locce, a tributary to the Belvedere glacier system, have been recorded.

On several occasions these floods damaged alpine pastures, ski lifts and houses above Macugnaga, a group of small villages that nowadays depends heavily on tourism as the major source of income. After a flood in 1979, large dams were constructed to control the main outlet stream from Ghiacciao del Belvedere. Swiss glaciologists have since co-operated with Italian scientists and civil defence authorities in monitoring this remarkable glacier, to the extent that it has become one of the most intensively studied glaciers in Europe. Despite being so well studied, the glacier produced an even bigger surprise in 2001. In June of that year, guardians of Rifugio Zamboni, a hut maintained by the Club Alpino Italiano, noticed that the lower part of Ghiacciao del Belvedere had started to swell up so dramatically that the ice had begun to tower above the nineteenth century moraine and in one place had even begun to over-ride it. The entire lower part of the glacier showed signs of strongly accelerated flow, such as intensive crevassing and partial detachment from its higher section in the Monte Rosa east face. The glacier was now experiencing what glaciologists somewhat cautiously called, a 'surge-like' stage. Instead of the usual 30 metres per year, the ice was moving at speeds of up to 200 metres per year (although the latter figure is rather slow for a 'real' surge).

Like many other glaciers, the advancing Nigårdsbreen, an outlet glacier from Josterdalsbreen ice cap, discharges water in an unpredictable manner. A sudden increase in discharge took these tourists by surprise, and they were lucky to escape with their lives.

As a result of the rapid ice motion, a roughly semi-circular depression formed on the ice surface at about 2150 metres above sea level that filled with meltwater. This supraglacial lake soon reached a surface area of 2500 square metres. The local authorities immediately realized the high risk of a major outburst of water. One year later, in summer 2002, the lake had grown to 150 000 square metres and contained approximately three million cubic metres of water. Parts of the village of Macugnaga were evacuated, as an outburst seemed imminent. Remarkably, the water level was so high that parts of the glacier were almost buoyant. However, despite extensive crevassing, the lake did not drain. One plausible explanation is that any drainage channels that might have developed were immediately closed because of the rapid motion of the glacier and internal deformation of the ice.

The Italian civil defence then staged a massive engineering operation to prevent an impending catastrophe. Hundreds of tonnes of technical equipment were airlifted by helicopters to the lake. Water pumps and pipes were installed in order to lower the lake level artificially, and a high-voltage power line was built all the way from Macugnaga to supply the electrical pumps with energy. This unprecedented and impressive effort was widely covered by Italian media, and soon many tourists started to invade Macugnaga to see the now famous 'lago ephemero', the temporary lake.

Visitors, however, were frustrated as access to the lake and construction site was severely restricted on safety grounds, as an uncontrolled lake outburst was still a possibility. Finally, by August 2002 the situation seemed to be under control and, as a result of pumping, the lake level had been lowered by several metres. In addition, the subglacial and englacial drainage system had become more efficient, allowing more water to drain from the lake naturally. Both 'lago ephemero' and the spectacularly advancing glacier, which by now had started to invade a pine forest, were soon the main tourist attraction of Macugnaga. Hundreds of people arrived every day to witness this unique glaciological spectacle.

At the time of writing, we do not know how the glacier will evolve. However, it is already clear that it serves as an excellent example of the impact of glaciological processes on mountain communities. As glacial outburst floods have the most far-reaching effects of all glaciological phenomena, it is imperative that such events are closely monitored, so that settlements many kilometres away from the glacier termini may not be endangered or damaged.

Floods in other areas

On a worldwide basis, outbursts from glacial lakes and of water from within glaciers are more frequent and widespread events than catastrophic ice avalanches. Repeated and catastrophic floods have been recorded since 1788 in the Mendoza valley in the Argentinian Andes, and since 1600 in the Oetztal in the Austrian Alps. In both cases a glacier surged out of a tributary valley and entered the main valley, damming the river and creating a lake. Once the water had built up to a critical level, it flowed over or breached the ice dam. Glacier surges in the Mendoza valley recur about every half century, each time causing a flood; the latest was in 1985. On the other hand, the glacier Vernagtferner, responsible for the Austrian floods, has receded to such an extent that no surges have been observed in the last hundred years.

14 Living and travelling on glaciers

Life and movement on glaciers offer a number of challenges that are outside the experience of most people. Solutions depend on whether the activity is in the accumulation area, and so prone to build-up of snow, or in the ablation area, where melting down of the ice surface may be a problem. In this chapter we first describe the glaciological aspects of an early overland journey to the South Pole, as this gives a flavour of what it is like to travel across the world's largest ice mass. We then take a more thematic approach by looking at the dangers of glacier travel, and then review the various modes of transport including walking, skiing, dog-sledging, motorized travel and flying. Lastly, we explore the ways in which people live on the glacier surface today, whether it be simple camping or in semi-permanent research stations.

A classic polar journey

The 'Heroic Age of Polar Exploration' of the period 1890–1914 saw the first serious attempts to explore the Antarctic Ice Sheet and assess its scientific significance. One of the most stunning journeys was undertaken by Sir Ernest Shackleton's 1908–1909 British expedition. One of Shackleton's goals was to reach the then unconquered South Pole, and he actually pioneered what has become the principal overland route to the Pole. This route crossed the Ross Ice Shelf, ascended a newly discovered trunk glacier that slices its way through the Transantarctic Mountains, and reached the polar plateau at over 3000 metres in the heart of the East Antarctic Ice Sheet. Shackleton's party of four members was forced to turn back because of insufficient supplies less than 150 kilometres from the Pole, in latitude 88° 23′ S. Quotes from Shackleton's diary published in his book *The Heart of the Antarctic* (William Heinemann, London, 1911) reveal

Camping on glaciers is one of life's special attractions to an Antarctic scientist, despite its sometimes arduous nature. Here, a geologists' camp of the US Antarctic Program is set up on the lateral moraines of Shackleton glacier at 85° S. Traditional-style Scott pyramid tents are supplemented by modern lightweight mountaineering tents. Flags are used as markers in case of blizzards.

the varying nature of the ice cover over which his party traversed, and the glaciological and meteorological hazards that had to be overcome, on this epic journey.

The long slog across the Ross Ice Shelf, then known as the Great Ice Barrier, characterized the first major leg of their journey – a flat, featureless plain with only glimpses of the Transantarctic Mountains on good days to relieve the boredom.

November 17. A dull day when we started at 9.50 a.m., but the mountains abeam were in sight till noon. The weather then became completely overcast, and the light most difficult to steer in; a dead white wall was what we seemed to be marching to, and there was no direct light to cast even the faintest shadow on the sastrugi [wind-carved snow dunes]. . . . Our march for the day was 16 miles 200 yards (statute) through a bad surface, the ponies sinking in up to their hocks. This soft surface is similar to that we experienced last trip south, for the snow had a crust easily broken through and about 6 in. down an air-space, then similar crusts and air-spaces in layers. . . . Today we had a plus temperature for the first time since leaving – plus 9° Fahr. At noon, and plus 5° Fahr. At 6 p.m. The pall of cloud no doubt acts as a blanket, and so we were warm, too warm in fact for marching.

Finding a route through the Transantarctic Mountains was the key to reaching the Polar Plateau. They chose the huge 250-kilometre long, 30- to 50-kilometre-wide Great Glacier that Shackleton subsequently named the Beardmore as their route. This glacier turned out to be far from easy; conditions ranged from heavy crevassing to hard, wind-polished ice. The debilitating wind and sun, and inadequate food took their toll. The two following quotes from the Lower Beardmore Glacier give a vivid picture of the glaciological hazards that the party encountered.

December 7. Started at 8 a.m., Adams, Marshall and self pulling one sledge. Wild leading Socks [the sole surviving pony] behind. We travelled up and down slopes with very deep snow, into which Socks sank up to his belly, and we plunged in and out continuously, making it very trying work. Passed several crevasses on our right hand and could see more to the left. The light became bad at 1 p.m., when we camped for lunch, and it was hard to see the crevasses, as most were more or less snow covered. After lunch the light was better, and as we marched along we were congratulating ourselves upon it when suddenly we heard a shout of 'help' from Wild. We stopped at once and rushed to his assistance, and saw the pony sledge with the forward end down a crevasses and Wild reaching out from the side of the gulf grasping the sledge. No sign of the pony. We soon got up to Wild, and he scrambled out of the dangerous position, but poor Socks had gone. Wild had a miraculous escape. He was following in our tracks, and we had passed over a crevasse which was entirely covered with snow, but the weight of the pony broke through the snow crust and in a second all was over. Wild says he just felt a sort of rushing wind, the leading rope was

Crevasses are one of the main obstacles to glacier travel. If they are visible, as is this one on Griesgletscher, Canton Valais, Switzerland, then deviations from one's route are necessary. If they are bridged by snow, then roping-up is necessary, as is constant probing if one is to avoid falling through a bridge.

snatched from his hand, and he put out his arms and just caught the further edge of the chasm. Fortunately for Wild and us, Socks' weight snapped the swingle-tree of the sledge, so it was saved, though the upper bearer was broken. We lay on our stomachs and looked over into the gulf, but no sound or sign came to us; a black bottomless pit it seemed to be. . . . When we tried to camp to-night we stuck our ice-axes into the snow to see whether there were any more hidden crevasses, and everywhere the axes went through. It would have been folly to have pitched our camp in that place as we might easily have dropped through during the night. We had to retreat a quarter of a mile to pitch the tent. It was very unpleasant to turn back, even for this short distance, but on this job one must expect reverses.

The party soon passed onto a section dominated by bare glacier ice, but continuing with a maze of crevasses. Bare, highly wind-polished ice is a feature of some of the major trunk glaciers through the Transantarctic Mountains, and indicates that there is no net accumulation in these regions, in contrast to the Ross Ice Shelf below and the Polar Plateau above. The party was now climbing steadily uphill, and relaying proved necessary at times. However, their spirits were uplifted by the splendid mountain scenery and good weather.

December 10. Falls, bruises, cut shins, crevasses, razor-edged ice, and a heavy upward pull have made up the sum of the day's trials, but there has been a measure of compensation in the wonderful scenery, the marvellous rocks and the covering of a distance of 11 miles 860 yards towards our goal. We started at 7.30 a.m. amongst crevasses, but soon got out of them and pulled up a long slope of snow. Our altitude was 3250 ft. above sea-level. Then we slid down a blue ice slope, after crossing crevasses. Marshall and I each went down one. We lunched at 1 p.m. and started at 2 p.m. up a long ridge by the side moraine of the glacier. It was heavy work, as the ice was split and presented knife-like edges between the cracks, and there were also some crevasses. Adams got into one. The going was terribly heavy, as the sledge brought up against the ice-edges every now and then, and then there was a struggle to get them started again. We changed our foot gear, substituting ski-boots for the finnesko, but nevertheless had many painful falls on the treacherous blue ice, cutting our hands and shins. We are all much bruised. We camped on a patch of snow by the land at 6 p.m. The rocks of the moraine are remarkable, being of every hue and description.

Although not trained in geology, Shackleton gives a good description of the rocks in the moraine. That evening he still had the energy to climb about 200 metres up the adjacent mountain, a prominent landmark they named The Cloudmaker, in order to collect *in situ* rock specimens. He also noted that 'the glacier is evidently moving very slowly, and not filling as much of the valley as it did at some previous date, for the old moraines lie higher up in terraces'. This perceptive observation indicates that Shackleton was aware that the ice sheet must have once been much thicker. It is now known that the moraines he described are glacial deposits several million years old.

By 18 December, Shackleton's party had reached the edge of the polar plateau, after a huge effort through varied terrain of hard ice, moraine and variable snow, always climbing steadily, yet noting the landscape and geology as they went. Here they make their last depot, and thereafter travel light with just the clothes they were wearing, one tent and reduced rations. The Pole was still 500 kilometres away. The lower temperatures (28° of frost [Fahrenheit] on Midsummer's Day, 20 December) and wind, combined with the still increasing altitude and short rations, left them perpetually cold and hungry. Day after day they struggled onwards, as the following typifies.

January 5. To-day head wind and drift again, with 50° of frost, and a terrible surface. We have been marching through 8 in. of snow, covering sharp sastrugi, which plays havoc with our feet, but we have done 13⅓ geographical miles, for we increased our food, seeing that it was absolutely necessary to do this to enable us to accomplish anything. . . . Our temperatures at 5 a.m. were 94° Fahr [negative]. We got away at 7 a.m. sharp and marched till noon, then from 1 p.m. sharp till 6 p.m. All being in one tent makes our camp work slower, for we are so cramped for room, and we get up at 4.40 a.m. so as to get away by 7 a.m. Two of us have to stand outside the tent at night until things are squared up inside, and we find it cold work. Hunger grips us hard, and the food-supply is very small. My head gives me great trouble. I began by thinking that my worst enemy had it instead of myself, but now I don't wish even my worst enemy to have such a headache; still, it is no use talking about it. Self is a subject that most of us are fluent on. We find the utmost difficulty in carrying through the day, and we can only go for two or three more days. Never once has the temperature been above zero [Fahrenheit] since we got on the plateau, though this is the height of summer. We have done our best, and we thank God for having allowed us to get so far.

January 7 and 8 saw them pinned in their tents during a blizzard, leaving just January 9 to get as far south as they could, the latest date on which they still had a chance of a safe return. At 88° 23′ S, 162° E they hoisted the Union Jack presented by the Queen, and 'took possession of the plateau in the name of His Majesty'.

Retracing their steps, after the disappointment of turning back, was an equally arduous task. They made rapid progress down the Beardmore Glacier, despite frostbitten feet and numerous falls into crevasses (although in harnesses), while on the Ross Ice Shelf they were all struck down with dysentery from eating horsemeat and were held up by occasional blizzards, but they all returned safely, by the skin of their teeth, meeting the ship finally on 4 March.

Dangers of glacier travel

Before venturing onto a glacier, the visitor has to evaluate the dangers that might be encountered. The seriousness of the hazard depends on whether the ice surface is snow-free, snow-covered or debris-covered. Imagine we are walking up a typical valley glacier in summer in the Alps, heading for the summit of one of the 4000-metre peaks. As we approach the snout we enter a zone of rapidly changing water courses and loose, saturated debris that is prone to mass movement, especially as mud-flows. The ease with which we can set foot on the glacier depends on whether the glacier is receding or advancing. Receding glaciers have gentle snouts, and the majority of glaciers today are in this state. If, however, the glacier is advancing, the snout will be near vertical and possibly heavily crevassed. It will then be necessary to attach crampons to one's boots, cut steps with an ice axe and also, if the terrain is particularly difficult, to rope up.

Having gained access to the glacier surface it is likely that we will find sets of crevasses. Since we are in the ablation area, the crevasses are not snow-covered and are therefore obvious, and we would normally bypass them. Indeed, melting back of crevasses often means that we can walk in and out of them, although this is fairly arduous and crampons will be needed.

By knowing something about glacier flow we can predict where the heaviest crevassing will occur. It is particularly intense at the margins, where a glacier flows over a bedrock step (an icefall) or round a bend. By heading for the middle of the glacier we can often gain smoother, almost crevasse-free ice. Walking on such terrain may be both very easy and enjoyable, particularly if the sun has been shining for several days. Intense solar radiation makes the ice surface quite rough, and boots find a good grip without crampons. On the other hand, if it has rained recently, the ice will have melted uniformly and become very slippery.

Meltstream on the surface of White Glacier, Axel Heiberg Island, Canadian Arctic, beginning to emerge from the winter snow cover. Streams on Arctic glaciers can be several metres deep and be un-crossable, especially when the snow is beginning to melt and large areas of slush develop beneath the snow cover. The slush can sluice off down-glacier with little warning.

Crossing glacial meltwater streams may be more difficult in the afternoon than in the morning due to changing amounts of discharge. The upper picture was taken below Vadret da Morteratsch on a July morning, the lower in the afternoon after ablation and subsequent runoff had increased considerably.

Moving up-glacier, we may next be faced with an icefall, and the chances are that it will stretch right across the glacier. Therefore, to continue further will necessitate more crampon work, step-cutting and roping up. An icefall typically has séracs, remnant towers between crevasses. These pose a serious hazard as they can topple without warning, so speed is of the essence.

Still in the ablation area, we may find that large parts of the glacier are debris-covered. Blocks, typically several metres high, litter the surface of the glacier. Block size relates to how well bedded or jointed the rock is; thus granite and gneiss blocks are large, but sedimentary rocks are typically small. Although thick debris slows down melting, the underlying ice continually changes its form, so loose debris on the surface is constantly shifting and sliding. Extreme care is needed when crossing such terrain. Crampons are of little use, and if used will get damaged. These areas are also associated with steep ice faces, typically 10 metres high, which are smeared with debris, the wetness giving a dark appearance to the face. It is easy to overlook these ice faces. Stepping on them invariably leads to an uncontrolled slide to the bottom, and since ponds are commonly developed in hollows, a cold freezing dip may be the consequence.

If the ablation area is crevasse-free, another obstacle that might be encountered is a melt-water stream. On alpine glaciers, a stream is typically incised to depths of one or two metres (and often much more on Arctic glaciers). It is usually deep in relation to its width. Crossing it may require a long detour. In fact, it may be more beneficial to walk *down*-stream, despite the stream getting bigger. This is because streams commonly disappear into moulins – large vertical shafts by way of which water reaches the bed.

If we are walking near the glacier margin we have to contend with other hazards. Rock, ice and snow avalanches may travel down onto the ice surface. With recession and lowering of the glacier surface since the Little Ice Age, lateral moraines are also unstable, and the loose material within them is constantly falling onto the ice surface, albeit in only small volumes at a time.

As our trip continues, we eventually approach the snowline. Clean ice is now giving way to granular firn that progressively becomes wetter until we are in a zone of slush. This is usually a narrow zone on a temperate valley glacier. In contrast, in Arctic regions the slush zone may be an impassable snow swamp. Since the slush can suddenly become mobile on slopes, care is required to avoid disturbing it unduly. We have now reached the accumulation area where snow covers a variety of holes, notably crevasses. At this point roping-up is essential. The purpose of this is that if one

On the highland icefields of Spitsbergen, Norwegian Arctic, the weather can change suddenly. This storm approaching Wilsonbreen arrived within a few hours, and blizzard conditions subsequently prevailed for three days, confining the party in their tents.

person falls through a snow bridge, then the others should hold the fall (as described below).

Continuing upwards we may be faced with yet another icefall. Here, the snow cover on the one hand may hinder safe crossing through the crevasses and séracs, but on the other can facilitate passage as big holes might be bridged. Such zones can be death traps. Many people have been killed, for example, on the notorious Khumbu Icefall on Mount Everest, mainly as a result of collapsing séracs. Unfortunately, this icefall provides the only reasonable access to the higher camps on the Nepalese side of the world's tallest mountain, en route to the summit. As the ground becomes steeper towards the head of the glacier we may be faced with an exceptionally large crevasse called a bergschrund. This cleft separates ice adhering to the mountain face, from the actively moving glacier. Snow bridges may be the only way of gaining access to the higher slopes of our mountain.

Considering all the obstacles and dangers described so far, a trip on a glacier may not seem to be particularly enjoyable. However, as glaciologists we have become used to avoiding trouble whenever possible. Rather, we have come to appreciate the myriad of colours of ice and rock, the multitude of ablation forms, as well as the gurgling sounds of melt-water streams.

A typical day in the Alps may start off with fine weather, but by the afternoon clouds commonly build up, and our descent may have to be made in poor visibility. In view of this, recognizing key glaciological features on the way up is of value in finding one's way down. Crevasses may give a clue as to the position of the margins, whilst following longitudinal foliation or medial moraines, which form parallel to the flow direction, is a good way of keeping one's bearing in thick mist. As we walk off the end of the glacier, we have to be sure to be on the correct side of the outlet stream, as this will have grown dramatically during the day as a result of increased ice and snowmelt.

Acclimatization

Since most glaciers occur at high altitude or high geographical latitude, visitors need to spend time adapting to the conditions. At low elevations in the Polar Regions this is simply a matter of drinking enough fluids and keeping warm, but not allowing the body to sweat excessively. Antarctic first aid manuals give detailed instructions on staying healthy, and especially avoiding hypothermia. Overheated Antarctic bases are not conducive to good acclimatization for outside fieldwork, and people adapt much better to the cold when camping or staying in primitive huts.

In high mountain regions it is acclimatization to altitude that is the main health concern. Most lowland people adapt fairly rapidly to medium elevations, say of around 1500–2500 metres. However, above this level gaining height needs to be undertaken at a steady gentle pace if one wants to avoid the worst effects of altitude sickness. For example, on the well known trek in Nepal from Lukla airport at 2700 metres to Everest base camp on Khumbu Glacier at 5200 metres, at least 10 days is recommended for gaining this altitude. Thus, the day-to-day increases in altitude are only of the order of 300–600 metres. Two or three rest days at intermediate levels are also suggested. Even then trekkers and climbers may experience the early signs of acute altitude sickness: they lose appetite, suffer from severe headaches, experience vomiting, become tired very easily, wake up from sudden shortages of breath, find it hard to sleep and cough a lot. Altitude sickness is the result of lack of oxygen. Over half the visitors to elevations above 3500 metres suffer from some kind of mountain sickness. After a while these symptoms become less severe. If the problem does not go away it is best to descend a few hundred metres.

Too rapid an ascent, especially above 4500 metres, can result in life-threatening illnesses, including the lung-related high-altitude pulmonary oedema or cerebral oedema which affects the brain. Both can have a negative effect on bloodflow and hence insufficient oxygen is delivered to the brain. The former is characterized by a sudden decrease in performance, coughing and foamy and bloody saliva. The symptoms of the latter are extreme headaches, confusion

Walking on glaciers is fraught with many dangers, and roping-up is usually necessary where the hazards are hidden. Here, during a training exercise, roped parties are crossing crevassed terrain on Ross Island, with the glacier-covered slopes of Mount Terror in the background, whilst the flat snow-covered area below is the McMurdo Ice Shelf.

Cross-country skiing with a fibreglass 'pulk' (lightweight, boat-shaped sledge) on Lomonosovfonna, a highland icefield in Spitsbergen. This is an efficient way of travelling over the undulating, snow-covered glaciers.

and hallucinations. In both cases death may occur within hours, but fortunately can be prevented by descending rapidly to a lower altitude. This, however, may not be possible for climbers on difficult ascents. Many mountaineers have thus perished at high elevations.

Above about 5300 metres the human body does not adapt at all. This is not surprising, since at this altitude there is only half the air pressure compared with at sea level. Climbers who are aiming for the summits of the highest peaks need to be already well acclimatized at this level. Many of them will use oxygen cylinders for the ascent. Since the body will deteriorate above this altitude rapidly, experienced and strong climbers try to reach the summit quickly, and return to base camp after spending no more than three or four days at higher camps. For scientists working on a glacier tongue at altitudes between 4000 and 5000 metres level, acclimatization is possible over a period of a month or so. However, we have found that work efficiency is much less than at low altitudes, and beyond this time most of us have become tired out and are keen to return home.

Modes of glacier transport

Glacier techniques

Although traversing a glacier can be as safe as any other outdoor activity, as long as one can understand and read the processes going on, it is important to be aware of the alpine mountaineering techniques available in order to maintain safety. As this book is not a manual on how to behave on a glacier, we only give the briefest of summaries here, and the reader is referred to the Bibliography for further information. Key things to consider when venturing on a long glacier journey on foot are physical and mental fitness, technical capability and ability to cope with an emergency, such as a fall into a crevasse.

Warm clothing is essential on any mountaineering trip, but with the expectation that temperatures on the ice could be several degrees lower. Boots should be rigid or semi-rigid so that crampons can be attached. Wearing crampons one needs to walk with feet wide apart, otherwise one stumbles as the crampon spikes become entangled in trousers. Although the ice surface is often rough enough not to need crampons, a spell of rain can soon make the surface slippery and treacherous. In addition, as a minimum, the glacier walker will need an ice axe for cutting steps, probing crevasses, self-arrest on steep snow slopes and maintaining balance when jumping obstacles such as streams. If venturing into the accumulation area, where crevasses are bridged by unstable snow, a range of technical mountaineering equipment and experience in using it is needed. This equipment includes ropes, harnesses, karabiners, ice axes and other devices. Members of a roped party all need to be able to hold a companion if he or she falls into a crevasse, whether in front, in the middle or behind – and be able to get that person out by setting up belays and pulley systems. The ability to cope with such an event requires serious training beforehand.

Ski-touring on glaciers

Travelling on snow-covered glaciers is easier on skis. Furthermore, a person's weight is more evenly spread over the snow, reducing the chances of breaking into a crevasse. On steep alpine terrain specially designed mountaineering skis are favoured. These have bindings that enable the heel to be raised on ascending a slope, while synthetic skins are fixed to the underside to allow forward glide, but offering resistance in the reverse direction. For downhill travel, the heel locks into the base plate, the skins are removed, and the skis behave much like normal downhill skis.

Some skiers prefer the much lighter langlauf (cross-country) skis. They are especially suited to travel on long, gently graded glaciers, as distances can then be covered rapidly. The simple bindings may allow the heel to be fixed to the base plate for downhill travel, but overall they offer less control and less security than mountaineering skis when descending steep slopes.

Dog-sledging

The tradition of dog-sledging stems from the indigenous peoples of the Arctic, the Inuit, who have relied on this technique to traverse sea ice, snowfields and occasionally glaciers in search of food since at least 50 BC. The Inuit have relied on the Greenland husky for a millennium, having long recognized its special qualities of patience, endurance and love of hard work. The most successful polar explorers of the so-called Heroic Age in the late nineteenth and early twentieth centuries purchased Greenland huskies to support their travel. Dogs proved to be particularly adept at hauling heavy sledge loads across ice shelves, up outlet glaciers and across the Polar Plateau to the South Pole, as demonstrated by the Norwegian, Roald Amundsen, on his successful expedition in 1912. Starting with 51 dogs and finishing with 14, he minimized weight by killing the weakest dogs and feeding their meat to other dogs. He was able to complete his dash for the South Pole and back, a distance of nearly 3000 kilometres, in just 89 days.

Huskies became the mainstay of Antarctic exploration throughout much of the twentieth century. They were used to support glaciological investigations and geological surveys by the Falklands Islands Dependencies Survey (later British Antarctic Survey) for half a

A British Antarctic Survey dog-team on Shambles Glacier, Adelaide Island, Antarctica. The person in front is checking out the terrain for hidden crevasses. For much of the twentieth century dog-sledging was the essential mode of transport for scientists working in the 'deep field'. With the withdrawal of dogs from Antarctica under environmental regulations in the mid 1990s, travellers now have to rely on mechanized transport. (Photograph courtesy of Nick Cox.)

The strongly built Greenland husky was the mainstay of Antarctic dog-sledging. This animal, outside Scott Base in McMurdo Sound, was a member of the last New Zealand dog-team, shortly before its removal to New Zealand in 1987.

century from 1945 until their withdrawal in 1994. During this period 108 dogs were imported and 850 born on the continent. The animals were bred in the Antarctic, and detailed records kept so that inter-breeding was minimized. By the end the husky was uniquely adapted to its role in support of Antarctic research. It is estimated that the cumulative distance covered by Survey dog teams was over half a million kilometres, most of which was on and around the Antarctic Peninsula. The longest unsupported journey undertaken was 1120 kilometres and lasted 67 days, but latterly the teams were taken into the deep field in Twin Otter aircraft. A typical team consisted of nine dogs arranged in two modes: one with a centre trace with paired dogs and a leader, and the other a fan trace where each dog is on an individual line. The latter system resembled the Inuit approach, and was preferred on crevassed glaciers because dogs could be rescued individually if they fell through a snow bridge.

Even when snowmobiles replaced dog teams for most serious fieldwork activity, dogs were kept for recreational reasons. It was said that on a small base, in the depths of winter, when tensions between the occupants became unbearable, an individual could reduce their stress levels by going outside and communing with the dogs.

Dog teams gave added security when travelling in Antarctica because, on a crevassed glacier, the driver was normally on the back of the sledge, several metres behind the leading dog. If any dogs fell through a snow bridge, the driver could stop and haul them out. Some explorers even claimed that dogs were able to 'sniff out' crevasses. However, this is something of a myth, and occasionally the driver would go ahead of the team in order to probe for crevasses.

During summertime travelling, huskies were normally left outside, on traces far enough apart to prevent the otherwise inevitable fights from breaking out. Despite their aggression towards each other, when they could almost fight to the death, the huskies could be very affectionate and gentle with their masters. The dogs were often fed seal meat, after which they were quite happy lying on the snow. During a blizzard they would curl up in a tight ball and allow themselves to become buried by snow. The work for the dogs could be exceptionally hard, but there seems little doubt that for most of the time they really enjoyed it.

Sadly for devotees of traditional polar travel, dogs were removed from Antarctica progressively from 1991 under the terms of an environmental protocol which forbade the introduction of alien animal species (except humans of course). No dogs were permitted after April 1994 on the grounds that animals such as seals could catch viruses like distemper, and that seals had to be culled to feed them. Many polar workers felt these arguments were weak, especially when motorized transport is hardly environmentally friendly in view of the pollutants it emits.

With the demise of Antarctic dog-sledging, there are few areas of the world where glacier travelling by this mode survives. What remains is mainly for the benefit of tourists, such as in Svalbard or on Jungfraujoch, high in the Swiss Alps.

Man-hauling

Although a well-trained dog team and driver was the most efficient mode of travelling over glaciers before mechanization, man-hauling on skis was equally popular with the early British explorers. Both Scott and Shackleton experimented with Siberian ponies and huskies, but resorted to man-hauling for the push to the South Pole. The men strapped themselves into uncomfortable canvas harnesses and dragged their sledges laden with food and camping equipment.

Long distances could only be achieved with animal support in the early stages of the march, together with the laying out of depots. Of all the activities on polar glaciers, man-hauling was by far the most arduous. Jerking a sledge into motion was said to feel like having your insides crushed against your backbone. Not only did these explorers have to contend with deep soft snow and the ever-present danger of falling into crevasses, but also with hard snow that had the texture of sand, and bare wind-polished glacier ice, as Shackleton's journey, described above, demonstrates. Cherry-Garrard, one of

Scott's companions, wrote 'Polar exploration is one of the cleanest and most isolated ways of having a bad time that has ever been devised'.

Man-hauling also was an important mode of glacier travel in the Arctic. Many expeditions in Svalbard and Greenland used this method over the interior glaciers for geological exploration. The simplicity of travel compensated for the complexities of dog-sledging, but at the expense of back-breaking work and burning a vast amount of calories. In the days prior to mechanized transport, the typical sledge was approximately three to four metres long, and based on a design by the Norwegian explorer, Fridtjof Nansen. Made of ash and bound by twine, the sledge would flex readily over obstacles, whilst any broken stay could be easily replaced. The same type of sledge was used for dog-sledging, but had the addition of a rear platform on which the driver could ride.

Man-hauling is an activity pursued by many adventurers and scientists today, but rarely to the extent of the back-breaking work of the past. Glaciologists, for example, will use purpose-built sledges to drag around ice-drilling or radar equipment. Mountaineers wear comfortable harnesses attached via rigid poles to traverse glaciers; this approach is particularly efficient when using skis.

Mechanized transport

For undertaking scientific work in glacierized terrain, the traditional approaches of dog-sledging and man-hauling have been largely superseded by snowmobiles (also known as skidoos or snow-scooters). As the names suggest, they resemble a motorcycle, but have one or two skis at the front and tracked propulsion behind. The wide variety of models on the market today cater for all tastes, but glaciologists generally need a heavy-duty reliable workhorse, rather than the more popular, fast, highly tuned vehicles that most people favour today. Snowmobiles are commonly used in tandem with a sledge that is usually constructed of aluminium. Passengers may be carried on the sledge, or alternatively towed on skis behind, a technique known in Norway as skijoring.

In recent years, some Antarctic operators have favoured the use of small four-wheel drive all-terrain vehicles for individual transport requirements. These powerful machines can cope with a wide variety of surfaces, ranging from large boulders in moraines to hard snow, and even bare glacier ice if the tyres are fitted with studs.

Larger vehicles have also been designed for hauling heavy loads over snow. In fact, the earliest use of a snow tractor was the experimental vehicle used by Scott in Antarctica in 1911. However, despite the sterling efforts of the mechanic this vehicle was a failure. Large specially designed 'Sno-cats', as they were called, were used by the Commonwealth Transantarctic Expedition led by Sir Vivian Fuchs in 1956. They not only successfully accomplished the first crossing of the Antarctic Ice Sheet, but provided enclosed accommodation in which to sleep and perform seismic experiments to measure the depth of the ice sheet for the first time. Modified Ferguson farm trac-

For travel in small parties snow-scooters ('skidoos') are the ideal mode of transport over snow-covered glaciers. Here members of a geological party pause beneath 1500-metre-high peaks during a six-week summer-time traverse of the icefields of Ny Friesland in northeast Spitsbergen.

tors were used at this time, notably by Sir Edmund Hillary's supporting party, but these were much less comfortable than Sno-cats.

Nowadays, a wide range of large vehicles are used in Antarctica, many of which are the modified equivalents of machines used for over-snow transport or to prepare ski pistes, as in the Alps, Scandinavia or Rocky Mountains. Some are powerful enough to transport prefabricated huts or tow several huge sledges, such as those that regularly move large quantities of material for ice-core drilling from the Russian station of Mirnyy to Vostok.

Over-snow transport in the Polar Regions is complemented by air support, especially by ski-equipped aeroplanes and helicopters. The US military in co-operation with the Royal New Zealand Air Force provide logistical support to McMurdo Station and Scott Base, using the neighbouring ice shelf as a landing strip. Outside the dark winter months Hercules and Starlifter transport planes fly frequently from

In mixed terrain of bare ground and snow-free glaciers, all-terrain vehicles have become popular. Here a geological party has just parked its Honda ATVs at camp at Hamilton Point on James Ross Island, just as a blizzard begins. Icebergs in Admiralty Sound can be seen through the driving snow.

Christchurch. The versatile Hercules aircraft also supplies the American's Amundsen-Scott Station at the South Pole, and serves other major deep-field operations such as ice-drilling or geological camps.

Smaller fixed-wing aircraft such as the De Havilland Twin Otter are widely used to support field operations in the Arctic as well as Antarctica. These aircraft have short take-off and landing capability, and can land on surprisingly bad surfaces if necessary. Light aircraft are popular in some glacierized mountain regions for tourism and mountaineering. In the Southern Alps of New Zealand glacier landings by Cessna aircraft are very popular, whilst in Alaska and the Yukon, mountaineers are airlifted to the foot of their peaks.

Helicopters have also proved their worth in support of science and tourism on glaciers. Although most helicopters have limited range, they have the advantage of being able to land almost anywhere. In Antarctica, some government organizations, such as those of the

For carrying heavy equipment and large numbers of people in Antarctica tractor trains are commonly used. Here we see a Kiwi field party on the sea ice in front of Barnes Glacier cliff on Ross Island. This versatile Swedish-built tracked vehicle, a Hägglund, can even float if it breaks through the sea ice.

USA and Australia, keep a fleet of helicopters on station to support fieldwork, and sometimes set up advance bases to support large-scale remote field activity. Another approach is that adopted by Germany and the Royal Navy (on behalf of the British Antarctic Survey), and also Australia, which is to use ship-based helicopters. As always with helicopters in Antarctica, it is customary to operate in pairs in case of mechanical failure. Helicopters are also popular with the tourist industry, providing scenic glacier flights in places such as New Zealand and the European Alps.

The development of reliable, safe mechanized transport has revolutionized access to glacierized high mountains and the interior of the polar ice sheets in the last 20 years. Although there have been many fatal aircraft accidents in glacierized terrain, the worst have involved commercial flights. One of the strangest incidents was the disappearance of a commercial flight from Buenos Aries to Santiago in South America on 2 August 1947. The aircraft was a converted Lancaster bomber and had five crew and six passengers on board. It disappeared just a few minutes from landing, but despite extensive searches no trace of it was found. It was only after aircraft wreckage and bodies started emerging from the ablation areas of a glacier on the slopes of the 6800-metre peak of Tupangato in the Andes in early 2000 that the explanation slowly emerged. It transpired that the aircraft, perhaps unknowingly slowed down by the Jet Stream and

At the top end of the mechanical transport spectrum is this huge comfortable bus, named 'Ivan the Terrabus', presumably after one of the Russian Tsars. It is used by the US Antarctic Program to ferry passengers to Williams Field, the runway on McMurdo Ice Shelf near Ross Island.

affected by poor visibility, began its descent to Santiago too early and crashed into the upper reaches of a steep glacier, bringing down with it an avalanche that buried the wreckage. As this was in the accumulation area, the plane took an englacial path through the glacier, before re-emerging in the ablation area at a much lower altitude.

Another tragic event was the loss of a DC10 from New Zealand on a scenic flight in Antarctic in November 1979. Because of a navigation error and poor visibility, the aircraft ploughed into the ice-covered slopes of Mount Erebus on full power and all 257 people on board died. Fortunately, nowadays, aircraft accidents are few, and the main hazards remain on the ground, especially when people are driving snowmobiles in poor visibility through crevassed terrain.

Camping on glaciers

Despite its arduous nature, camping on glaciers is often a highlight for the field-orientated scientist. Getting out of one's sleeping bag

Major scientific operations in Antarctica that involve ice drilling or co-ordinated geological projects require the installation of the equivalent of a small village. For this purpose the US Antarctic Program relies on LC-130 ski-equipped 'Hercules' aircraft, as here in support of a geological programme on Shackleton Glacier in the Transantarctic Mountains.

on a sunny morning, when snow crystals sparkling in the low sun provide a foreground to impressive peaks, is a delightful experience. A glaciological field party may need to spend several weeks at the same spot on a glacier, while geologists investigating bedrock in glacierized mountain country may find that glaciers provide the only flat ground on which to pitch their tents. Camping is most straightforward above the snow line. Tents with wide valances are needed because pegs are useless. In addition to equipment, snow blocks, cut with a specially designed saw, can be piled onto the valances (snow flaps) to weigh down the tent. There are disadvantages to camping on snow, however. Melting snow for drinking water and cooking requires a lot of energy, and using the typical pressurized camping stove is a very slow process. Also, during blizzards, equipment may be buried by snowdrifts so it is important to place it in a staked-out line to minimize digging out. Going out in blizzards is a hazardous and unpleasant venture as the whiteout conditions remove all sense of the difference between ground and sky, and it is easy to become disorientated and fall over. One has to be prepared for blizzards lasting several days in the Polar Regions or at high altitude.

Camping on snow when melting is underway can be very unpleasant. Snow turns to slush and everything becomes soaked, so such

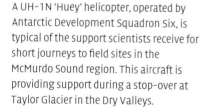

A UH-1N 'Huey' helicopter, operated by Antarctic Development Squadron Six, is typical of the support scientists receive for short journeys to field sites in the McMurdo Sound region. This aircraft is providing support during a stop-over at Taylor Glacier in the Dry Valleys.

Camping equipment in Antarctica must be able to withstand hurricane-force winds, and the most widely used tent is the pyramid, little changed from the version used by Scott and others a century ago. Pyramids are anchored down by rocks, snow and supplies. Here we see a French-manufactured 'Squirrel' helicopter approaching a camp on Shackleton Glacier, Antarctica with a view to resupplying a US geological party.

areas are best avoided. Down in the ablation area the problems are different. Melt-water may be abundant, but tents have to be secured with boulders. With time, the tent may end up on a mound of ice as it retards the process of ice melting, forming an unusual type of glacier table. Frequent repitching may be necessary, especially when tents become vulnerable to wind. For this reason, glaciologists will tend to camp on moraines as the melting here is slower, while stones can serve as temporary shelters and anchors for tents and equipment.

A big challenge associated with camping on glaciers is waste disposal. In the past rubbish and human waste was buried in deep holes or shoved down crevasses or moulins, while surplus equipment has just been left lying on the ground. Certain areas resemble rubbish dumps, the worst being places like the icy upper slopes of Mount Everest where discarded oxygen canisters and climbing equipment litter the landscape. This is no longer acceptable in today's environmentally aware society, and stricter controls are necessary to prevent further degradation of the environment.

Tight regulations are now an integral part of Antarctic operations.

Only icebreakers can sail safely through iceberg-infested waters, and even then underwater iceberg projections ('keels') can pose a serious hazard. Here the Royal Navy icebreaker *HMS Endurance* is undertaking a hydrographic survey between James Ross Island and Snow Hill Island in the Antarctic Peninsula region, so maintaining a grid pattern of necessity, which entails sailing close to clusters of icebergs.

Under the Environmental Protocol of 1994, all countries are working towards removal of all rubbish and solid human waste from the continent. Several countries already strictly implement the regulations. Thus the ideal mode of operating in glacierized terrain, whether remote or accessible, is to leave no impact that would be visible a year later.

Temporary scientific research camps

Large-scale operations, notably ice-coring, have been a feature of many glacier studies, from the polar ice sheets to the tropical ice caps. The infrastructure requires supply by transport aircraft. Once established, the stations need to operate for several years, and be self-sufficient in winter, serving the needs of dozens of scientists and support staff. Buildings are prefabricated and of modular construction. Energy is supplied by diesel generators, supplemented by solar and wind power. Strict environmental controls in places such as Antarctica, Greenland and the Alps mean that all materials need to be removed at the end of the project. The clean-up operations can be just as expensive as setting up the project in the first place.

Moveable buildings such as Jamesway huts provide heated comfortable accommodation including showers, although the outside

An unusual luxury for a British Antarctic Survey glaciological party in the 'deep field', several weeks into a field season, is a hot bath, improvized from a hot-water drilling system for measuring internal deformation of an ice stream. (Photography courtesy of David Vaughan.)

toilet facilities, being a hole excavated in the glacier, are somewhat primitive.

'Permanent' stations

Long-term monitoring of environmental change is a feature of several scientific stations in Antarctica. For determining atmospheric and climatic changes, for example, records spanning as many years as possible are necessary. Several stations were established on the Antarctic Ice Sheet during the International Geophysical Year in 1957, and have operated continuously ever since then.

The American South Pole Station is located on the Polar Plateau and has been rebuilt several times, because the buildings buckle as they become buried by snow. Supply of this remotest of stations is by fixed-wing aircraft, but plans are underway to bulldoze a surface road from McMurdo Station on the coast, 1600 kilometres away. Completion is planned for the 2004–2005 season. Crevasses will be opened with explosives and snow bulldozed to fill them. The

A geological field party camping in the accumulation area of Wilsonbreen, an outlet glacier in northeast Spitsbergen, with the peak of Backlundtoppen (1081 metres) in the background. Tents are secured by piling snow on the valences. The vehicle to the left is an all-terrain 'Argocat', fitted with snow tracks. The party had to be completely self-sufficient for a period of six weeks as it traversed this remote glacierized terrain.

German Georg von Neumayer Station on the Ekstroem Ice Shelf has also become buried, so to get to it the occupants have to drive down an underground tunnel cut out of the snow. Sub-surface accommodation is warm and comfortable, although some staff find the absence of daylight difficult to get used to. Being near the edge of the ice shelf, it is supplied mainly by ship, with tractors running across the shelf to the station. Transport facilities and fuel are kept at the surface, but require frequent excavation from snow drifts. If a station has become buried like this it is difficult to remove the debris when it is no longer useable, but ultimately it will be transported seawards out of sight before being discharged into the sea embedded in an iceberg.

The Russian station of Mirnyy is built on top of a grounded ice wall, again in an area of net accumulation. Having been relatively unconcerned about environmental issues, the station now has rubbish forming part of the snow stratigraphy in the coastal cliff, whilst nearby is a graveyard of wrecked aircraft half buried in the snow.

Permanent stations on the Antarctic Ice Sheet have developed a variety of strategies to cope with snow accumulation, especially as snow drift tends to bury any object. The simplest approach is to use reinforced buildings and gradually allow them to be buried, such as Germany's Georg von Neumayer station on the Ekstroem Ice Shelf. Living underground, however, is not the best way of experiencing Antarctica.

Major scientific programmes in Antarctica, away from the main bases, require complex logistical requirements. American operations are usually large-scale, as this geological expedition on the Shackleton Glacier in the central Transantarctic Mountains indicates. The camp was supplied by ski-equipped Hercules aircraft, with more local transport provided by two 'Squirrel' helicopters and a fixed-wing Twin Otter. The buildings on the right are heated 'Jamesway' huts for living quarters, while scientists in transit use the pyramid tents on the left. In front of the helicopter is a fuel bladder to supply the aircraft. The area, occupied for one summer season, was thoroughly cleaned after use and all waste removed.

With their Halley Station on the Brunt Ice Shelf, where the ozone hole was discovered, the British have adopted a quite different approach to construction in an area of net snow accumulation. The building stands on legs that can be jacked up as necessary to keep it clear of the snow. Wind blowing beneath it keeps the area from drifting up. Located close to the ice shelf edge, this station is also supplied by sea. However, because there is a natural snow ramp from sea ice at the bottom of the ice cliff, offloading of supplies is easier than at Georg von Neumayer.

Permanent stations on ice are a feature only of Antarctica. Large resources are available to government-sponsored Antarctic organizations operating within an international context, a benefit not available to other scientific operations elsewhere. Of course, in Alpine regions, there is not the same necessity to build permanent stations when centres of population are close at hand.

15 Earth's glacial record

Many people are aware that during the last couple of million years the Earth experienced a series of ice ages, when ice sheets expanded into regions now occupied by major cities such as Chicago, New York, Birmingham, Oslo, Stockholm, Berlin, Moscow and Zürich. However, this is only part of the story because the geological record shows that for over 3000 million years (m.y.), Earth's history has been dominated by warm 'greenhouse' conditions, punctuated by a series of cold phases or 'ice-house' conditions when extensive glaciers and ice sheets developed.

For the last cold phase on the Earth, the Cenozoic Era, the evidence for extensive ice cover is usually in the form of a wide variety of landforms (as described in Chapter 10), and soft poorly sorted sediment called **till**, while associated sand and gravel are indicative of deposition from glacial meltwaters. Evidence for glaciations in the more distant past is sketchier, but is represented mostly in the form of ancient glacial deposits, commonly referred to as **tillites** (till which has hardened into rock), and associated sediments and erosional phenomena.

The purpose of this chapter is to outline how ideas concerning ice ages developed, how geologists recognize the evidence for glaciation in the rock record, provide a summary of when and where the main periods of glaciation took place, and to make some observations on the causes of glaciation. An understanding of ice ages throughout the past 3000 m.y. of Earth history can give us considerable insight into the workings of the climatic system. Information derived mainly from old rocks is complemented by the high-resolution records that can be obtained for the Cenozoic Era, notably sediments recovered from the deep oceans and continental shelves, and ice cores from the two remaining ice sheets and other smaller ice masses. We can use our knowledge of past ice ages to reconstruct former ice ages, and to

The legacy of glaciation in regions that lost their glaciers after the last glaciation 12 000 years ago includes some of the world's finest landscapes. Crummock Water, with the Central Fells of the Lake District beyond, is one of the classic glaciated areas in Britain, and one appreciated by many visitors.

There are several lines of evidence for glaciation in ancient rocks. One of the most useful is the presence of rocks mixed with particles ranging from clay to boulder size. This example is at Tillit Nunatak in East Greenland, and is of Late Proterozoic age (about 650 m.y. old). Tillit is Danish for 'tillite' the name give to rocks of glacial origin.

Figure 15.1 The geological time-scale, indicating when the Earth was subjected to continental-scale ice ages.

Million Years	Era	Period	Major glacial phases
0	Cenozoic	Quaternary	Glacial / interglacial cycles in N.Hemisphere ⎤
1.8		Neogene	⎬ Cenozoic ice ages
23.8		Palaeogene	Antarctic Ice Sheet ⎦
65	Mesozoic	Cretaceous	
142		Jurassic	
206		Triassic	
248		Permian	
290	Palaeozoic	Carboniferous	Permo-Carboniferous ice ages of Gondwana
354		Devonian	
417		Silurian	
443		Ordovician	Ordovician / Silurian ice age of Gondwana
495		Cambrian	
545	Pre-cambrian	Proterozoic Eon	Global ice ages? ('Snowball Earth'); Late Proterozoic
2500			Early Proterozoic ice ages
		Archaean Eon	[Oldest rocks, 3800m.y., in Greenland]
4600			

help predict future trends in glacier and ice sheet growth or decay. To provide a framework for understanding Earth history, geologists have divided time, in descending hierarchical order, into Eons, Eras, Periods and smaller divisions. The table illustrates the major time divisions and summarizes when the ice ages occurred, although we only describe those events that were of a continental scale.

The development of ice age concepts

The 'Ice Age' theory developed in the early nineteenth century, and accounted for many of the loose, widely scattered deposits at the Earth's surface, subsequently regarded as Quaternary in age. Although the famous German poet Goethe has been given credit for first suggesting the concept of ice ages, the leading protagonist of the theory was the Swiss natural historian Louis Agassiz, who had worked extensively on modern glaciers in the Alps. The application of this new concept to the older geological record soon led to recognition of Permian glacial deposits in India and Australia in 1859, and South Africa in 1870. The first Precambrian glacial deposits were described from Scotland in 1871 and Norway in 1891, the latter site becoming famous for its well-developed striated pavement. Since then ancient glacial deposits of many ages have been recognized throughout the world. Unravelling the nature of ice ages remains one of the most important scientific debates today; the clues that ice ages yield are fundamental to assessing the future climatic prospects of our planet.

Recognizing the signs of glaciation

Scientists recognize that the processes occurring in glacial environments produce one of the most complex associations of sediment on the planet. This is because there are various ways in which debris is transported by a glacier, but also because of the interaction with a range of non-glacial processes, such as mass-flow, river action, marine and lake currents and wind. Earlier chapters in this book describe these processes and the products that can be found in the modern landscape record. If we can identify these features in the geological record we are able confirm the former presence of glaciers.

Erosion surfaces commonly provide the most compelling evidence of glaciation. These, however, are normally not well preserved over long periods of time. Large-scale forms such as ancient valleys may occasionally be found, although equally useful, but smaller, abrasional forms such as grooves and striated surfaces are more common. Chattermarks and crescentic gouges are often associated with such surfaces. Grooves may be produced by meltwater under high pressure, and this can lead to a suite of regular, deeply incised bedrock channels and less regular forms. Intermediate-scale erosional forms, which combine evidence of rock fracture and abrasion, are typified by roches moutonnées. Exhumed ancient examples are also known.

Glacial erosional landforms, such as these roches moutonnées in the Sahara desert, Mauritania, are rarely preserved in the geological record, but where found they are excellent indicators of former ice-flow directions, in this case from from left to right. These are exhumed examples dating back to Late Proterozoic time.

Rocks of glacial origin (tillites), if they have not been tectonically deformed, sometimes weather in such a way that the stones can be removed easily. If the stones have been glacially transported, as has this example from Late Proterozoic strata in Mauritania, they may retain striations, indicating abrasion as they were transported on the underside of a sliding glacier.

A variety of depositional features enable us to determine whether sediment is of glacial origin or not. In other words, is the sediment a till? According to some geologists, 'till' is the most varied of all geological materials under a single name. The characteristic sediment released from the base of a glacier is one that contains everything from clay to boulder size, but these proportions vary tremendously. Near their point of origin stones may be dominant (80 per cent or more of the total sediment), whereas in the most distal parts reached by the ice the till may be almost all clay with less than one per cent stones. Regardless of the proportion, stones in the deposit will have a variety of shapes as a result of partial rounding in contact with the bed. The stones may also show striations or chattermarks. Supraglacial sediment, in contrast, is made of more angular and coarser-grained fragments. The two types of sediment may be juxtaposed when the ice melts.

Terrestrial glacial sedimentary environments include deposits produced by glacial rivers (glaciofluvial sediments). These sediments contain well-sorted sand and gravel, as well as a variety of bedding structures. Reworking of glacial sediment by proglacial streams may even remove all clear signs of glacial deposition. We

Evidence of glaciation from 2000-2500 m.y. ago is shown by this 'tillite' – a mixture of coarse and fine fragments in a very hard rock, Whitefish Falls, Ontario, Canada.

One of the advantages of examining old rocks of glacial origin is that they are commonly well exposed. On Ella Ø in central East Greenland, folding of the Late Proterozoic glacial rocks has tilted them up on end, so one can follow a stream course and work progressively up or down through several hundred metres of glacial strata. These rocks bear many of the hallmarks that have led to the generation of the 'Snowball Earth' hypothesis of global glaciation.

know this because the modern glaciofluvial-dominated flood plains in the European and New Zealand Alps show few signs of glacial transport.

As glaciers recede, over-deepened basins emerge, which progressively fill with glaciomarine or glaciolacustrine sediments; such deposits are commonly well preserved in the geological record.

Apart from till deposited at the ice margin, and sand and gravel brought into the water body by subglacial streams, there are also laminated sediments that settle out from suspension. We can recognize a glacial context by the presence of ice-rafted dropstones, which fall into the laminated sediment and deform it.

Lastly, we may mention other evidence of cold climates that may be preserved in the rock record, although this too is often enigmatic. Wedge-shaped polygonal features penetrating bedding surfaces, such as sand in till, may be interpreted as ice-wedge casts of periglacial origin. Another feature, although rare, is the presence of silt, which may represent wind-blown glacial dust called **loess**.

Earth's ancient glacial record

It seems that, as the Earth has evolved through geological time, warm climates have been the norm, whereas glacial conditions were the exception. Nevertheless ice ages clearly have occurred on many occasions in Earth's history. At least one ice age during the late Proterozoic Eon may have been global in extent.

Earliest evidence of glaciation

South Africa provides us with the earliest evidence of an ice age on Earth. Dating of the host rocks yields Archaean ages of around 3000 m.y. ago. When considering glacial periods in the far distant past we must bear in mind that continents had a totally different geographical configuration from those of today, one that so far remains unresolved.

Early Proterozoic (2500–2000 m.y. old) glaciations

Early Proterozoic rocks of North America, South Africa and Western Australia bear the first evidence of continental-scale glaciations. The most studied glacial rocks are about 2300 m.y. old and are widespread in Ontario, Canada. Here, a sequence of tillites is spectacularly exposed at Lake Huron, on surfaces abraded by the completely unconnected last great ice sheet of only 10 000–20 000 years ago.

China has abundant evidence of Late Proterozoic glaciations, spanning at least three time intervals. Not only are there are extensive tillites, but there are grooved and striated pavements underlying them, as here in Henan Province.

Dropstone structures are among the most telling indicators of former glaciation, in this case Late Proterozoic glaciation in Namibia. Dropstones are formed when icebergs release debris into laminated sediments of the sea or lake floor. These dropstones near Narachaamspos in the Kaokoveld have been used to promote the 'Snowball Earth' hypothesis as a global refrigeration event, while the overlying pink and orange carbonate rocks are indicative of the succeeding hot phase.

Similar rocks are found in mid-central USA, Finland, South Africa and Western Australia spanning the interval 2500–2000 m.y. ago.

Late Proterozoic (1000–540 m.y. old) glaciations

After the Early Proterozoic glacial era there appears to have been several hundred million years without any glacial activity. This interval may have been relatively warm as a result of the extensive extrusion of lavas, combined with out-gassing of carbon dioxide leading to the development of a global 'greenhouse' condition. However, from around 1000 m.y. to early Cambrian time (about 540 m.y. ago) Earth was affected by a number of ice ages, at least one of which was on a scale greater than at any other time in Earth's history. The scale of these Late Proterozoic events was so enormous that some scientists have argued that the Earth completely froze over. This is the so-called Snowball Earth scenario that has captured the attention of the media and popular science authors.

Late Proterozoic tillites occur on every continent, and magnetic data from the rocks, used to determine the latitude at the time of deposition, indicate near-equatorial glaciation for many occurrences. Several ice ages occurred within this 400 million year timespan, however, and reconstructing the distribution of the tillites is not possible with any accuracy because of imprecise dating.

Of all the major glacial eras, the Late Proterozoic has always been the most controversial. It took detailed studies of modern glaciers, followed by application of newly learnt principles of glacial deposition to the controversial deposits to reaffirm a glacial influence. The 'Snowball Earth' hypothesis invokes almost complete icing over of the planet, including all the oceans, following major chemical changes in the atmosphere. This cold period may have come to an abrupt end. The disappearance of the ice cover in just a short time interval is attributed to a runaway greenhouse effect following a massive release of carbon dioxide during volcanic eruptions.

There are other plausible explanations for Late Proterozoic ice ages, including uplift of major mountain ranges, or changing of the angle of the Earth's axis of spin. Whatever the ultimate explanation, there is no doubt that the debate has stimulated much valuable research in determining the origins of ice ages.

Early Palaeozoic (570–410 m.y. old) glaciation

The early part of the Palaeozoic Era is represented by the Cambrian Period that, apart from a few local occurrences, has no glacial sediments. Indeed, some scientists link the explosion of life at this time to the disappearance of the Late Proterozoic ice sheets, and the onset of conditions favourable to the evolution of many life forms. Conditions unfavourable to the development of large ice sheets persisted until the end of the Ordovician Period. Then, around 510 m.y. ago, a new ice age affected those continents that were located over the South Polar Region, including Africa, the Arabian Peninsula,

Among the best-preserved striated and grooved pavements in the world are those of Permo-Carboniferous age (around 290 m.y. ago) in South Africa. Although long-buried by younger sedimentary strata, in places, such as near Douglas in the Great Karoo region, the pavements are being exhumed, preserving features that are as fresh as the day they were formed.

Evidence of another glaciation in the Kaokoveld, a hot desert area of northern Namibia. Here, a Permian–Carboniferous landscape has been preserved for most of its life by younger sediments. These have now been exhumed, revealing classic glacial landforms such as this glacial trough at Omutirapo.

Europe, and North and South America. However, it is unlikely that extensive continental ice sheets existed for more than one or two million years.

Perhaps the most impressive evidence of Ordovician glacial activity comes from the central Sahara Desert, where a wide range of features of continental glacial erosion and deposition, and periglacial environments are represented. Geologists are reasonably sure that all major land-masses were then united in a single 'supercontinent' called Gondwana.

Late Palaeozoic (410–235 m.y. old) glaciations

After the Ordovician-Silurian ice sheets had disappeared, a period of almost ice-free conditions prevailed on Earth until the initiation of a series of Carboniferous-Permian ice ages. This period, the best-known of all the ancient ice ages, lasted approximately 100 m.y. Massive ice sheet sheets waxed and waned over Gondwana (Africa, Antarctica, Southern Asia, Australia and South America).

Africa, once again, has the best-preserved evidence of glaciation. The Karoo Basin of southern Africa is famous for its exceptionally well-preserved glacial erosional and depositional features, while in northern Namibia a glacial landscape of troughs and cirques has been exhumed, revealing a freshness of form that is remarkable.

Late Palaeozoic ice ages have been linked to active tectonic movements and uplift associated with plate margins and inland sedimentary basins. Not all areas were glaciated at the same time, an observation that is linked to the migration of the Gondwana supercontinent over the South Polar Region. Thus, glaciation took place mainly in the Carboniferous-Period in South America and the Permian in Australia, with peak glacial activity straddling the Carboniferous/Permian boundary. Some geologists have suggested that no single ice sheet developed, but rather that several ice masses waxed and waned at different times across Gondwana. However, it is equally plausible that, for much of the time, a large ice sheet straddled the various continents at least near the Carboniferous-Permian boundary.

An interesting aspect of Late Palaeozoic ice ages is their influence on global sea levels and, indirectly, on the industrial development of the western world. The sedimentary sequences in Britain, Germany and the USA, which carry the Carboniferous Coal Measures on which the Industrial Revolution was built, were strongly influenced by sea-level fluctuations driven by the waxing and waning of the Gondwana ice sheets. When the ice sheets were large, sea-level was

The long-term glacial record, spanning some 35 million years, has been obtained by drilling in the deep ocean from research vessels, or on the continental shelf by rigs designed for drilling on land, but placed on the winter sea ice. Here the seven-nation Cape Roberts Project is drilling through approximately 2 metres of sea ice into the seafloor in the western Ross Sea. Over a kilometre of core was obtained over three seasons. (Photograph courtesy of Peter Barrett.)

low and coal-generating swamps developed, but when the ice sheets waned sea level rose, flooding the land and triggering the formation of limestone. So this major phase of glaciation has played a significant role in the formation of two vitally important economic resources, coal and limestone.

The 'greenhouse' world of the Mesozoic Era (235–65 m.y. ago)

Three periods make up the Mesozoic Era (the age of dinosaurs): the Triassic, Jurassic and Cretaceous. A few places show equivocal signs of glaciation but, on the whole, the Era was characterized by the warmest climates since the Proterozoic Eon. The typical deposits are red sandstones of the type formed in hot tropical deserts today, and chalk and mud characteristic of warm shallow seas. However, by Cretaceous time global cooling was underway, but it was to be many millions of years before extensive glaciers and ice sheets developed once again on Earth, in the succeeding Cenozoic Era.

Earth's Cenozoic glacial history (35 m.y. to present)

For a long time, the Quaternary Period (marking the last Period within the Cenozoic Era) was regarded as synonymous with glaciation. This period represents the last two million years of Earth's history. However, with advances in offshore drilling technology and the acquisition of long sediment cores in the early 1970s, scientists were surprised to find that, in Antarctica at least, glaciation had begun over 30 million years earlier! We now know much more about this phase of Earth history than about all the earlier glacial periods put together. The late Cenozoic Era coincides with the evolution of the human race and, as time went on, ice ages had an increasing impact on the survival and migration of our species.

Nature of the evidence

Huge fluctuations of the northern hemisphere ice sheets took place during the Quaternary Period, leaving an imprint on 30 per cent of the Earth's land surface. The nature of the evidence for glaciation is much more varied than for ancient glaciations. On land the record is usually sketchy, as later glaciations tended to remove the evidence of earlier ones. The wide range of sediments and landforms produced by glaciers, as described in Chapter 10, may be used to reconstruct the locations and fluctuations of former ice sheets and valley glaciers. For example, end-moraine complexes define the limits reached by a glacier. Commonly, there are several moraine systems, so if such features can be dated, then rates of recession can be determined.

Much of what we know about Cenozoic ice ages comes from the offshore record, where continuous sequences of sediment are preserved. Scientists in the early 1970s discovered that the deep-sea sedimentary record, recoverable by drilling ships, could be used to reconstruct glacial/interglacial climates. In order to determine the pattern of climatic change, a technique known as oxygen isotope analysis was developed. This technique involves determining the ratio between two varieties of oxygen, the light isotope ^{16}O and the heavy isotope ^{18}O. As ocean water evaporates, preferentially more ^{16}O is released, but in non-glacial times is returned almost immediately to the ocean as runoff from the land. In glacial times this excess ^{16}O is stored in ice masses, leading to enrichment of ^{18}O in the oceans. Marine sediment containing microfossils called foraminifera, which reflect the composition of seawater. As the sediment accumulates on the sea bed, a continuous record of oxygen isotopic variations is

Figure 15.2 The relationships between sea level, ice ages and oxygen isotopes. The left-hand diagram represents conditions during an interglacial period, and the right-hand diagram an ice age. During the ice age the sea level drops and an ice sheet grows. Since ^{16}O evaporates more readily than ^{18}O, the ice sheet becomes enriched with ^{16}O and the ocean enriched in ^{18}O. The sediments accumulating on the sea floor preserve the isotopic signature of the ocean. (Adapted from Bennett, M. R. & Glasser, N. F., *Glacial Geology*, Chichester: Wiley, 1996.)

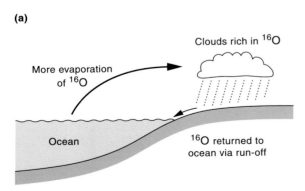

(a)

More evaporation of ^{16}O

Clouds rich in ^{16}O

Ocean

^{16}O returned to ocean via run-off

Oceans are not enriched in ^{18}O as ^{16}O returns relatively quickly to the oceans, thereby maintaining the balance of $^{18}O/^{16}O$

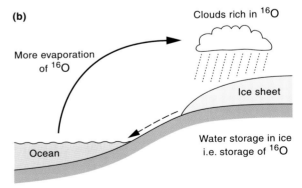

(b)

More evaporation of ^{16}O

Clouds rich in ^{16}O

Ice sheet

Ocean

Water storage in ice i.e. storage of ^{16}O

Ocean becomes enriched in ^{18}O as a result of storage of ^{16}O in ice sheets

The desire to know when Cenozoic glaciation began has stimulated extensive offshore drilling in Antarctica. The recovery of core, such as this example from New Zealand's CIROS-1 hole in McMurdo Sound, has illustrated the longevity of Antarctic glaciation (35 m.y. old), but a gradual transition from temperate to cold polar glaciation has taken place over this period, as manifested in the style of sedimentation and fossil remains. The number at top left indicates the depth below the sea floor from which the sample was obtained.

produced, with the highest values of ^{18}O occurring during interglacial periods. The main problem with the deep-sea record is that it often does not show any *direct* evidence of glaciation. So, for a comprehensive direct record of glaciation on the longer time scale (tens of millions of years), geologists have turned their attention to the continental shelf areas of Antarctica.

Oxygen-isotope records of ice cores can also be used alongside records from deep-sea sediments, as a means of estimating trends in global sea level. The approach is to determine the ^{16}O /^{18}O ratio of small samples of melted ice. For a high-resolution record spanning several of the last ice ages in the last half million years, glaciologists have drilled through various ice sheets and ice caps in order to retrieve ice cores. The most comprehensive records have come from the Greenland Ice Sheet and the East Antarctic Ice Sheet. In contrast to the oceanic sedimentary record, ice cores show the reverse isotopic pattern, in that the highest values of ^{18}O occur during ice ages and the lowest during interglacial periods.

The deep-sea record

Sediment cores that record climatic history have been recovered from many sites in the deep-ocean, from the Polar Regions to the tropics. Over the last three decades, sophisticated drilling ships, capable of drilling to depths of several kilometres below the sea floor and staffed by expert drillers and scientists, have gathered huge volumes of data, while thousands of samples from each cruise have been analysed in the laboratory. This work has revolutionized our understanding of climatic change during the Cenozoic Era (and earlier). With the advantage of continuous sedimentation in many parts of the deep ocean, all the climatic fluctuations determined by oxygen-isotope analysis can be added up. In the past 800 000 years we see a pattern of glacial/interglacial cycles each lasting about 100 000 years, and between 800 000 and a million years cycles of 40 000 years. These cycles represent mainly the waxing and waning of the northern hemisphere ice sheets. The change in cyclicity at 800 000 years was accompanied by intensification of these ice sheets. The isotope record also indicates numerous fluctuations before the northern hemisphere ice sheets existed, thereby documenting fluctuations of a much more dynamic Antarctic ice sheet than that of today. With so many glaciations evident from the deep-sea record, it becomes apparent that the terrestrial record is so incomplete that correlation from place-to-place on land is a real challenge. Another interesting feature of the deep-sea record is the nature of each glacial cycle when examined in detail. First there is a slow gradual build-up of ice to a maximum. Then rapid deglaciation follows, accompanied by sea-level rise.

Scientists can also determine relative sea level according to the balance between the heavy and light oxygen isotopes. For example, during the peak of the last ice age, around 20 000 years ago, sea level was 120 metres lower than at present. The 'lost' water from the oceans was then locked up in the huge ice sheets that covered much of North America and Eurasia. In contrast, sea level today is almost as high as at any time in the last half million years.

The continental shelf record in Antarctica

To determine ice-marginal fluctuations and the climates under which glaciers form, we need a direct record of glacial sedimentation, whether it be from deposits released at the base of an ice mass, or from ice-rafted debris. For this reason, geologists have undertaken drilling on continental shelves, areas that had been over-ridden by glacier ice. The first continental shelf drilling was undertaken by the American-led Deep Sea Drilling Project in 1973 in the Ross Sea in a ship called *Glomar Challenger*. Since then there have been several major drilling projects focusing on the glacial and climatic record. Among the most successful projects has been drilling from sea ice in the McMurdo Sound region by New Zealand teams in the 1980s. The most successful of these projects was the retrieval in 1987 of a 702-metre long core that extended the glacial record of Antarctica back to some 34 m.y. ago. The international consortium known as the Ocean Drilling Program also undertook successful drilling in the Prydz Bay region in 1988 and 2000, and the Antarctic Peninsula in 2001, using the drilling ship *Joides Resolution*. Then, the international Cape Roberts Project of 1997–1999 drilled a series of holes in the western Ross Sea, using the sea ice-based techniques developed by New Zealand teams earlier. The Cape Roberts Project, a multi-million dollar venture, has been the largest to date, involving 55 scientists from New Zealand, the USA, Italy, the UK, Australia and the Netherlands.

The most comprehensive record of glacial/interglacial cycles in the Antarctic is from the Western Ross Sea (including McMurdo Sound). Well over two kilometres of core have now been obtained, providing a record of continental shelf glaciation, unequalled anywhere in the world. Based on a wide range of studies, the evolution of the ice sheet follows an interesting pattern of change. The oldest sediments recovered, about 34 m.y. old, indicate a cool temperate climate, with glaciers extending into mixed forests at sea level, as in

The most recent drilling programme is named the Cape Roberts Project and is a multi-million dollar operation, involving 55 scientists, that provides a series of snapshots of how the Antarctic Ice Sheet has developed since its initiation, 35 m.y. ago. Scientists are here examining the core for the first time in the laboratory at McMurdo Station.

southern Chile or Alaska today. The period from 25 to 17 m.y. ago shows substantial cooling, with a stronger glacial influence and tundra vegetation. A gap in the record until three m.y. ago is then followed by the cold polar conditions we see today.

These results are important because they tell us how Antarctica could change in response to predicted levels of global warming. The smaller extent of glaciers recorded by these sediment cores equates with substantially higher sea levels. The challenge now is to determine how quickly the Antarctic Ice Sheet will respond to the unprecedented temperature rises predicted for the next 100 years.

The onshore record

In Antarctica, where Cenozoic glaciation began, the Transantarctic Mountains and Prince Charles Mountains have yielded evidence of glaciation in the form of tillites dating back to at least 10 m.y. ago.

There is evidence in Antarctica of a much warmer ice sheet than the present day, coming in the form of glacial sediments known as the Sirius Group or, as here, striated and grooved bedrock immediately beneath. This example is from Roberts Massif, Shackleton Glacier.

The tillites are preserved along the flanks of major glacial troughs cutting through the mountains. However, scientists have vociferously argued about the age of the deposits for many years. Suffice to say is that, because these deposits are associated with vegetation (notably a shrubby beech called *Nothofagus*), the climate was much less severe than today's; a tundra plant association grew along the flanks of the glaciers.

The Quaternary Period represents the last stages of Earth history. The terrestrial record shows only four ice ages, in comparison with the large number recorded in the marine record. The concept of so few glaciations comes from sediments and landforms (notably end-moraines). The main problem in comparing different parts of the world is that the sedimentary record is fragmentary and the sediments are difficult to date. The only certainty concerns the scale and broad dimensions of the last ice age, which peaked about 18000 to 20000 years ago. This ice age goes under the names of Weichselian in northern mainland Europe, Würm in the Alps, Wisconsinan in North America, Devensian in Britain and Midlandian in Ireland. However, earlier glaciations were more extensive. On land, these glaciations have left behind a veneer of sediment commonly several metres thick, notably till, sand and gravel. In addition, glaciers delivered much of the sediment to the sea, dumping it at the edge of the continental shelf. One of the biggest challenges for geologists interested in the timing of events is to match the onshore record with that offshore.

Ice core records

Whilst geologists have probed into 'deep time' by drilling into offshore sediments in Antarctica, glaciologists have cored into the ice sheets themselves. In so doing, the latter have acquired shorter, but continuous, climatic records approaching half a million years long. Glaciologists have also taken deep cores from the Greenland Ice Sheet, as well as shorter ones from ice caps in warmer regions, including the 5650-metre-high Quelccaya Ice Cap (14° S) in southern Peru, the Dunde Ice Cap on the Tibetan Plateau (38° N), a highland

Evidence for glaciation just a few million years ago in Antarctica is preserved in these laminated deposits of the Sirius Group, Bennett Platform, Shackleton Glacier, Transantarctic Mountains. The laminated sediments contain large stones (dropstones) rafted by icebergs into a lake.

icefield at 5300 metres on Mount Logan (60° N) in the Canadian St. Elias Mountains, and near the summit of Monte Rosa in the Alps. Ice-coring operations in these terrains are difficult and costly, but thousands of metres of core have been recovered, which have given us records from many of the climatic zones on Earth.

There are a number of practical reasons why scientists are investigating the atmospheric processes of the distant past. We urgently need accurate information about climatic evolution in order to establish the nature of, and reasons for, climatic changes today, as well as to predict future trends. It is a matter of great concern that we still cannot say with any certainty why and how ice ages begin and end; nor can we predict how the biggest single controller of rapid sea-level changes, the Antarctic Ice Sheet, will respond to climatic warming. However, the climatic record is preserved throughout the entire depth of certain parts of ice sheets and ice caps, where ice accumulation and glacier flow are slow, and melting is negligible. Within these ice masses relatively thin snow layers are deposited on top of each other annually and, apart from being slowly transformed into ice, remain undisturbed for thousands of years. After retrieving ice cores, the scientist analyses trapped air bubbles in order to decipher the atmosphere of the past. For example, the ratios of isotopes, especially heavy (^{18}O) and light (^{16}O) oxygen, and of hydrogen and deuterium, in ice indicate air temperatures at the time of deposition of the snow. In addition, carbon dioxide and methane in the bubbles indicate how the burning of fossil fuels has affected atmospheric composition since the onset of the Industrial Revolution.

A remarkably detailed three-kilometre-long ice core has been obtained from the Russian station of Vostok in the middle of the East Antarctic Ice Sheet, the place where the lowest temperature on Earth ($-89\,°C$) has been recorded. This core spans 420000 years and embraces four ice ages and interglacial periods. Collaboration, particularly between Russian and French scientists, has yielded a wide variety of data, notably CO_2 concentrations in trapped air and oxygen isotope ratios that are used to estimate sea-level change. During glaciations the CO_2 concentrations are low, whilst in interglacial times they are relatively high. Industrial pollution by 2000 had raised this figure to well above that of any previous interglacial period.

How do all these data fit together to provide a coherent story of climatic change? As the graph shows, the form of the CO_2 curve is very similar to that of the oxygen isotope curve. This indicates a strong correlation between CO_2 in the atmosphere and ice volume. Human civilization is thus running a dangerous experiment; by increasing CO_2 in the atmosphere we are in danger of causing significant reductions in ice volume.

The Greenland Ice Sheet has been the target of many ice-coring projects over the last half century. The earliest attempts, by a team of American scientists in 1956 and 1957, retrieved cores 305 and 411 metres long. As drilling techniques improved progressively greater depths were reached, and in the 1960s cores gave a remarkably detailed record of climatic evolution spanning more than 120000 years and embracing the last glacial/interglacial cycle.

The most recent cores are from two holes drilled near the ice divide in central Greenland within 30 kilometres of each other. The European Greenland Ice Core Project (GRIP) obtained a 3029-metre-long core between 1989 and 1992, while the USA's Greenland Ice Sheet Project (GISP2) had obtained 3053 metres of core by 1993. Both of these cores reached bedrock, and together provide an environmental record in excess of 100000 years. Oxygen isotope analy-

Figure 15.3 Vostok ice-core records for the past 420000 years. (a) Carbon dioxide levels. (b) Ice-volume records determined from oxygen isotope data. (c) Air temperature. (Data selected from Petit, J. R. et al., *Nature*, **39**, pp. 429–36), 1999.)

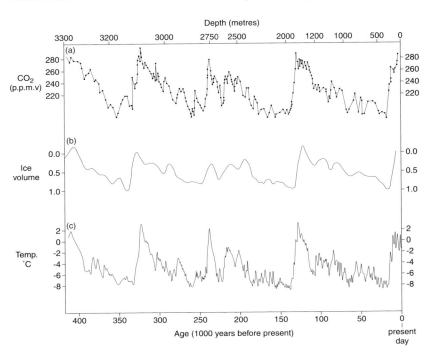

sis has provided the basic stratigraphy and record of climatic change, but many other physical and chemical properties have also been determined. The records from both cores are nearly identical, except in the lower 200 or 300 metres, where deformation (especially folding) has disrupted the layered sequence or stratigraphy in the ice. Interestingly, the oxygen isotope records also match closely that obtained from the Vostok core for the last glacial cycle, indicating that global events are being recorded.

The lower part of the Greenland cores represents the previous interglacial period and, although the details are difficult to unravel because of folding, it appears that rapid climatic fluctuations characterized this period around 110 000 years ago. From then until 15 000 years ago, the period representing the last ice age, the core records quite large temperature and snowfall variations with as many as 23 short-lived warmer phases or 'interstadials'. We then enter a phase of deglaciation, disrupted by a 1300-year-long cold phase called the Younger Dryas. This well-studied time interval was characterized by the reintroduction of glaciers to the British Isles for the last time, and expansions of existing glaciers elsewhere. The termination of the Younger Dryas approximately 12 000 years ago and transition into the interglacial period (the Holocene Epoch) in which we now live was abrupt; the temperature rose in Greenland by 7 °C and accumulation doubled in just a decade. This abrupt change supports the view that future temperature changes can be equally abrupt. The upper parts of the Greenland cores record many interesting changes in environmental conditions, including the Mediaeval Warm Period, the Little Ice Age (AD 1450–1900), and the impact of the Industrial Revolution.

Reconstructing former ice sheets

Mapping glacial sediments on the ground or from space provides the raw data that are needed to reconstruct former ice sheets. By adding glaciological parameters based on modern ice sheets, scientists can model numerically the size and behaviour of past ice sheets. These models can then be used to predict how future ice sheets

Glen Roy in the Grampian Highlands of Scotland is a famous nature reserve where the extent of past glaciation was noted in the late nineteenth century. Particularly significant are the three stripes across the hillsides, known as the Parallel Roads, that are notches cut by waves in a large ice-dammed lake, dating from around 12 000 years ago.

might behave. Models can predict ice thickness, as is the case of the last British ice sheet, or ice marginal recession over thousands of years in the case of the last Fennoscandian and Laurentide ice sheets.

Patterns and causes of ice ages

Ever since Ice Age theory was developed, scientists have sought to find the causes that result in the Earth switching between glacial and non-glacial modes. Some are related to the way the Earth moves around the Sun, whilst others are purely of a terrestrial nature.

Many scientists who have attempted to determine causes of climatic change have failed because they have had only an inadequate data-base to rely upon. The development of the concept of a large number of ice ages in the Quaternary Period is an illustration of this problem. In 1924, Milankovitch, following Croll in the late nineteenth century, startled the scientific community by announcing that variations in Earth's planetary motions resulted in cyclic variations in the nature of solar radiation input, and could be used to predict when ice ages might occur. Variations in Earth's planetary motions as proposed by Milankovitch include:

- Variations in Earth's axis of tilt, or the **obliquity of the ecliptic**, on a 41 000-year cycle.
- Change of shape or **eccentricity** of Earth's elliptical orbit in a 100 000- and 400 000-year cycle.
- Rotation of the Earth's spin axis as the planet orbits the Sun, or **precession of the equinoxes** on a 19 000- to 23 000-year cycle.

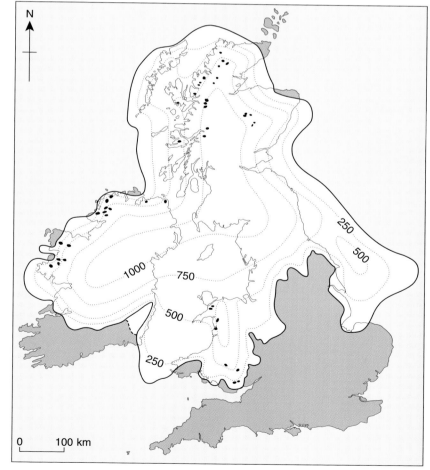

Figure 15.4 An ice-sheet reconstruction over the British Isles during the last glacial maximum, around 18 000 years ago. Calculated ice thicknesses in metres are also shown by the contours. The black dots represent nunataks, i.e. mountains projecting through the ice sheet. (Adapted from Boulton, G. S. *Geology of Scotland*. London: Geological Society of London, Chapter 15, Figure 15.16, 1991.)

Glaciers

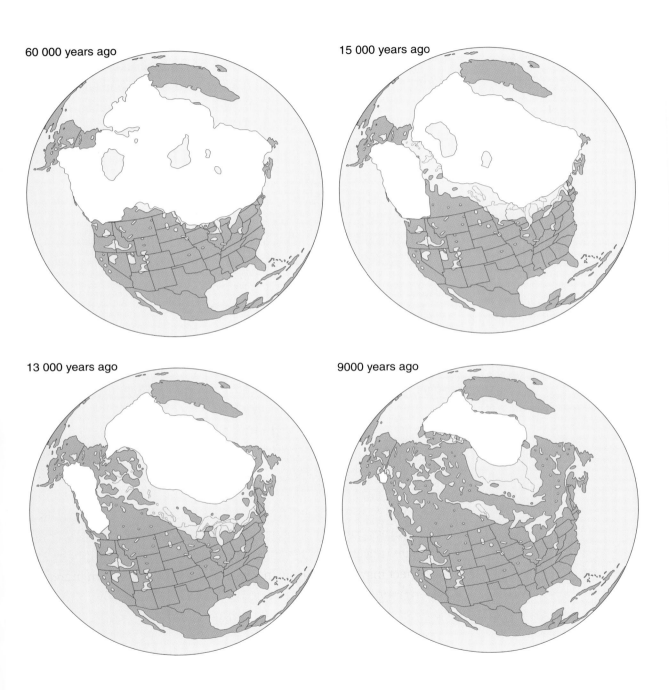

60 000 years ago

15 000 years ago

13 000 years ago

9000 years ago

Figure 15.5 Numerical modelling shows shrinkage of the Laurentide Ice Sheet over North America during the last ice age. Note how the Ice Sheet splits into two and how large lakes develop along the southern margin. Greenland at the top is excluded from the modelling. (Simplified from Marshall, S. J. & Clarke, G. K. C., *Quaternary Research*, **52**, pp. 300–15, 1999.)

Milankovitch's ideas were discredited at the time because geologists adhered slavishly to the concept of four Quaternary ice ages, for which fairly clear landform evidence was at hand from Southern Germany. However, no empirical data were available to support Milankovitch's hypothesis. Only in the last three decades have his ideas been largely vindicated. A 40 000-year cycle has indeed been detected in the deep-sea oxygen isotope records, that has lasted since about 23 m.y. ago. Oxygen isotope variations in deep-sea cores, which reflect global ice volumes, also closely match solar radiation curves. Climatic changes resulting from astronomical causes on a longer time-scale may have resulted from the increasing length of the day. Galactic motions may also have played a part in climatic change, such as when the Solar System passed through spiral arms of the galaxy.

Terrestrial influences on climatic change are numerous, but not all are likely to be significant. They include fluctuations of the Earth's magnetic field, variations in geothermal heat output, variations in composition of the atmosphere (especially of carbon dioxide), biological evolution and its influence on the carbon dioxide/oxygen balance, and the distribution of land and sea. The last includes the ratio of land area to sea, the degree of relief, the positions of the Earth's spin axis and the location of the poles in relation to land, the pattern of seaways and the extent to which land and sea areas are concentrated. Linked to the latter group of influences is the role that is played by tectonic uplift. Over a few million years uplift can alter the weather patterns dramatically. To give an example: if, as a result of plate movement, a continent drifts towards one of the poles, an ice sheet may start to grow in a highland area. This in turn increases the amount of sunlight reflected back into space, cooling the Earth further. In addition the polar continent will stop warm ocean currents reaching the polar region, and so on.

Throughout much of the geological record, the lack of the necessary precision in dating events prevents scientists from developing firmly based theories concerning the causes of ice ages. Nevertheless, we are in a position to make a reasonable judgement concerning the causes of some ice ages at least. We now draw together

The last full-scale glaciation in Britain of around 18 000 years ago transported large amounts of material southwards. The Isle of Arran in southwest Scotland had its own centre of ice dispersal, producing these whitish granite erratics that were then transported southwards down the coast a short distance and deposited on red sandstone bedrock.

the terrestrial evidence already noted in the preceding discussion, to determine the main cause of each of the glacial periods.

Little is known concerning the distribution of land and sea during the Archaean and Early Proterozoic glaciations, nor is there more than a very approximate idea of their ages or extents, or indeed how many occurred. Astronomical cycles have been considered as a cause, but there is little one can do other than speculate.

Deposits of the Late Proterozoic Glacial Era are now relatively well known, and from these several phases of ice ages may be identified. The present-day distribution of Late Proterozoic tillites indicates that glaciers expanded and contracted on a continental, if not a global, scale – a concept that has become known as the Snowball Earth hypothesis. We have to imagine an Earth with all land covered by ice and the sea frozen to depths of hundreds of metres – the ultimate deep-freeze. Determining the validity or otherwise of the 'Snowball Earth' concept is one of the most significant challenges faced by geologists interested in past glaciations. Two enigmatic characteristics stand out. The first is that on the basis of studies of the magnetic properties of the rocks many of the deposits were laid down in low latitudes (called palaeolatitudes). Some scientists have explained the low palaeolatitudes in terms of an increase in the obliquity of the ecliptic to more than 54°, which would have resulted in equatorial rather than polar glaciation. The second enigma is that the tillites are often intimately associated with allegedly warm-water carbonate rocks such as limestone. However, the deposition of carbonates need not necessarily be related to temperature alone, and in recent times geologists have demonstrated that at least some of the carbonates are of cold-water origin or have been reworked from older strata. One feature of the Late Proterozoic glaciations that we cannot ignore is their association with tectonically active areas. For example, the tillites of the North Atlantic region were associated with rift basins, the flanks of which were being uplifted. A modern analogue is the Transantarctic Mountain range on the edge of the Ross Sea rift basin. If uplift was a few thousand metres, this may have been sufficient to initiate large-scale glaciation.

The last glaciation in the United States saw the erosion of many fine glacial troughs, such as the Yosemite Valley in California. The over-steepened valley sides include El Capitan on the left, and the hanging valley with the Bridal Veil waterfall (dry) below on the right.

During the Palaeozoic Era two main periods with ice ages occurred, the short-lived Late Ordovician-Early Silurian glacial period and the much longer Carboniferous-Permian glacial era. Positioning over the South Polar region during both time intervals appears to have played a major role in initiating continental glaciation. The polar-positioning hypothesis, however, does not explain the abrupt changes of climate represented by more than one glacial epoch. For this we may also need to consider Milankovitch's planetary motion variables; these would help to explain the cyclicity in the Carboniferous Coal Measures in lower palaeolatitudes, the deposition of which was controlled by ice-influenced sea-level changes.

The onset of extensive Cenozoic glaciation began in Antarctica at least 35 m.y. ago, in response to a global cooling trend that began in the Cretaceous Period. Glaciation does not appear to be related to polar positioning directly because Antarctica appears to have been

Striking evidence of ice-sheet-scale glaciation can be found in Central Park, New York, where the exposed bedrock surfaces carry striations formed at the bed of a sliding ice mass.

centred over the pole for at least 100 million years. However, the separation of Antarctica from the other Gondwana continents, completed about this time, led to thermal isolation of the continent. In other words, once the circulation of ocean currents around Antarctica was unimpeded there ceased to be much transfer of heat from the other oceans. The uplift of the Transantarctic Mountains may have triggered glaciation about this time. The northern continents, where glaciation began much later, do not appear to have been influenced by their position relative to the pole. Cooling appears to have been initiated by tectonic activity whereby landmasses were displaced and oceanic circulation patterns changed. Then, once a certain threshold was reached, Milankovitch's variables resulted in glacial-interglacial cycles.

In conclusion we may ask the question: 'is there any evidence of periodicity of ice ages?' Many scientists have looked for regular cycles of ice ages over the Earth's entire history, which may be related to astronomical variables. Various cycles have been proposed, but all are based on inadequate data. The picture that emerges from recent studies is certainly one of *episodic* glaciation, but the evidence for *periodicity* over the entire span of geological time is lacking. In view of the apparent terrestrial causes of some of the glacial periods, this is not surprising. In contrast, on time-scales of around one million years or less, scientists have been able to identify cycles that are so short that continental movements are insignificant. This has certainly proved worthwhile for the Quaternary Period, which has yielded evidence for the cyclicity of glaciations related to astronomical variables on time-scales of about 40 000 and 100 000 years. The first indication of such cycles in the older record was found recently during the Cape Roberts Drilling Project in Antarctica described above. With so many scientists working on the core, the dating precision was far better than anything achieved previously! So, there is potential for identifying similar cycles in the older rock record, but in most places new techniques in dating are required to improve precision. Only then will there be scope for invoking astronomical parameters to explain climatic change in ancient glacial sequences.

16 Postscript: future prospects of glaciers

The study of glaciers is important for many reasons, not least because of the impact they will have on all our lives in the future. Perhaps most fundamentally, glaciers and ice sheets respond to climatic change. They affect global sea level, drive ocean currents and influence atmospheric circulation. Thus glaciers have an impact on human civilization from the poles to the tropics.

At a more local level, hazardous glaciers impinge directly on the lives of people in mountain regions, and some have been responsible for huge loss of life. We need to understand them better if future catastrophes are to be averted. Despite this, glaciers provide considerable benefits to human society, notably in terms of providing water and hydro-electric power, and more aesthetically as a focus of magnificent landscapes to be appreciated by mountaineers, skiers and other tourists. However, in the context of global warming and glacier recession, major concerns are arising about the loss of water supplies and tourist amenities.

Ice cores and glacial sediments provide unequalled records of climatic evolution, both within the span of our lifetime, and on timescales of millions of years. If we can establish the environmental record of the past, we will be better placed to understand the causes of climatic and sea-level changes, and thus be in a better position for determining what may happen in the future. Predicting the future behaviour of glaciers and ice sheets requires the application of many scientific disciplines. For example, the physicist can tell us about their dynamics, the climatologist about their state of health, the chemist about the relationship between atmospheric composition and glacier behaviour over thousands of years in the past, and the geologist about glacier fluctuations over hundreds to millions of years. The data produced by all these scientists are of vital importance to the numerical modeller, who tries to predict how glaciers

The vulnerability of the world's biggest mass of ice, the East Antarctic Ice Sheet, to climatic warming is less than for the peripheral regions. Indeed, warming may be heralded first by an increase in snowfall and a positive mass balance. Major glaciers draining the ice sheet through the Transantarctic Mountains will probably only respond slowly to climatic changes.

and ice sheets will respond to climatic change in the future. To reverse the old geological adage that 'the present is the key to the past', we can say that 'the past is the key to the future'.

What is happening to the Earth's climate?

In recent years there has been increasing recognition that human activities are changing our planet's atmosphere and climate. The Earth benefits naturally from a greenhouse effect whereby carbon dioxide and other gases trap solar radiation and prevent all the Sun's heat from re-radiating from the surface of the planet and back into space. However, most climatologists now agree that the global temperature is rising, and that this is at least partly due to the increasing levels of various gases, particularly carbon dioxide, in the atmos-

The terminus of Mackay Glacier, an outlet glacier from the East Antarctic Ice Sheet in the western Ross Sea, dwarfs a field party that is crossing the sea ice in front of it. The glacier has been slowly receding since its discovery in the early twentieth century.

phere, directly attributable to human activity. These increases arise from a combination of the burning of fossil fuels, which release carbon dioxide, and the destruction of forests, which convert carbon dioxide into oxygen. Increased levels of methane attributable to melting of permafrost and increasing numbers of domestic animals, together with the depletion of the ozone layer as a result of interaction with chlorofluorocarbons (CFCs) are also acknowledged as having a warming effect on the atmosphere.

In light of the above, there is growing concern amongst many scientists that the rising levels of pollution are exacerbating the greenhouse effect, and making the world's ice masses vulnerable to melting. According to the 2001 report of the Intergovernmental Panel on Climate Change (IPCC), global temperature rise from carbon dioxide emissions is set to rise between 1.4 °C and 5.8 °C by the year AD 2100. The increase is likely to be even greater in the Polar Regions where the largest ice masses occur. Indeed, for the worst-case scenario of uncontrolled emissions, predicted carbon dioxide levels and temperatures will reach levels not experienced by Earth at any time within the last 15 million years. Then, within 200 years, carbon dioxide levels will have reached concentrations not seen on the planet since 35 million years ago when there were no ice sheets at all. Such temperature rises are unprecedented in recent geological history. Given the volume of water locked up in glacier ice, the potential for the flooding of coastal regions in the next few generations is severe. If human civilization is to protect itself against such effects, emissions of pollutants must be curbed drastically. Following high-profile international conferences in Rio de Janeiro and Kyoto, in which the links between pollution and global warming were convincingly demonstrated, many governments have accepted that pollution from the burning of fossil fuels must be reduced and are taking steps to do so. Unfortunately, some governments, including that of the USA (the world's biggest polluter), refuse to accept the evidence and carry on as normal. Partly, the inability of governments to act is because of the unwillingness of the general public to accept increased taxes on fuel. 'Green' political parties in Europe

The signs of global warming in the form of glacier recession are present in most parts of the world. The biggest potential contribution to sea-level rise is the Antarctic Ice Sheet, but here the picture is more complex. In the northern Antarctic Peninsula, as here on the east coast of James Ross Island, we see not only the local glaciers thinning and disappearing rapidly, but also the collapse of ice shelves, producing numerous icebergs. Yet, elsewhere on the continent, the ice sheet appears to be growing.

promote the idea of increased taxes on petrol and diesel to support public transport, but mass public demonstrations against high taxes, as in Britain in 2001, have forced governments to backtrack. Development of cheap, efficient and convenient alternative forms of energy is urgently needed if the public is to reduce its dependence on fossil fuels.

Given the predicted increase in global temperatures, will the world's ice masses be subject to catastrophic meltdown or just slow gradual adjustment? So, to answer this question, let us now look specifically at how some of the world's ice masses will respond to global warming.

Impact of glaciers on global sea levels

There are various estimates available concerning how much water is locked up in the world's glaciers and ice sheets, and the effect this

Potential contribution of the world's ice masses to sea-level rise (U.S. Geological Survey 2001)

Ice masses	Metres of sea-level rise
West Antarctic Ice Sheet	8
East Antarctic Ice Sheet	65
Antarctic Peninsula	0.5
Greenland Ice Sheet	6.5
Mountain glaciers and ice caps	0.5

would have on sea level if it all entered the oceans. One of these estimates is given in the accompanying table. However, there are considerable errors in these figures, especially in relation to the Antarctic Ice Sheet, for which values ranging from 56 to 80 metres have been proposed in recent years. Nevertheless, these figures clearly demonstrate the potential impact on low-lying regions of the world even if only a small fraction of the ice cover melts.

Ice sheets

In Chapter 8 we have already seen the impact of regional warming on the ice shelves of the Antarctic Peninsula during the last 10 years. Several of them have completely disintegrated, although, because they were floating, their contribution to sea-level rise is negligible. However, is ice-shelf disintegration in the Peninsula a precursor to what might happen to the West Antarctic Ice Sheet? Some scientists have suggested that the West Antarctic Ice Sheet is potentially unstable, because much of it is grounded below sea level. It is only held in check by the buttressing effect of ice shelves, so if these were to disintegrate catastrophically, like those in the Peninsula, the thick grounded ice in the interior would rapidly discharge into the sea. This could lead to a global sea-level rise of several metres within a time-span sufficiently short to make it difficult for human beings to adapt. So is there any evidence to suggest that collapse of the West Antarctic Ice Sheet is imminent? Although there have been several major calving events from the two biggest ice shelves that hold back the ice sheet, the Ross and Ronne-Filchner (as described in Chapter 8), these may simply be part of the normal cycle of slow growth followed by large-scale calving on time-scales in excess of 100 years. Calving may simply be a response to the shelf becoming dynamically unstable, rather than to climatic warming.

Another feature of large ice shelves, such as the Ross, is that several ice streams feed them. Some flow at speeds of hundreds of metres a year, which glaciologists believe is related to the soft bed of sediment that deforms readily. Other ice streams have 'switched off', probably as a result of being frozen to the bed. On the other hand, one major ice stream in West Antarctica, Pine Island Glacier, is currently receding and thinning rapidly, and the basin that feeds it is also thinning. Thus, changes in the dynamics of the constituent parts of the West Antarctic Ice Sheet and its bordering ice shelves on

The pyramidal form of Nevado Santa Cruz (6247 metres), Cordillera Blanca, Peru viewed from the southwest at sunset. Although many ice masses in the tropical Andes are rapidly receding and it is predicted by some scientists that most will disappear in 10-15 years, the hanging glaciers on this peak are still looking healthy.

a time-scale of hundreds of years may be just as important as climatic change. This means that it is a major challenge for glaciologists to predict future behaviour of the ice sheet, and one that has not yet been resolved.

Most glaciologists believe that the East Antarctic Ice Sheet is much more stable than the West Antarctic Ice Sheet because most of it is grounded above sea level and is less likely to be influenced by the sea. Nevertheless, it has also been suggested that the ice sheet has surge-type basins, and hence could transmit large amounts of ice into the Southern Ocean. The Lambert Glacier system is one such candidate, although the evidence for former surge behaviour seems to us to be lacking.

Overall, then, there are major uncertainties about the future prospects of the Antarctic Ice Sheet. Apart from the northern peripheral areas, such as the Antarctic Peninsula, it is likely that warming

Glaciers are receding rapidly in the Rocky Mountains and many small ones have disappeared in the least 100 years. Larger glaciers, however, have many decades of life in them, when fed by snows of the highest peaks, such as Mount Robson, British Columbia, Canada.

will result in increased precipitation in the form of snow, and hence growth of the ice sheet. These two effects might cancel each other out and, according to some numerical models, it will take a rise of more than 5 °C to trigger meltdown.

The Greenland Ice Sheet is more susceptible to climatic change than the Antarctic Ice Sheet. Again there is much uncertainty concerning its future prospects under a global warming scenario. There does, however, seem to be a consensus that over the next hundred years the peripheral areas of the ice sheet will continue to recede as the equilibrium line rises, while the interior will thicken as a result of increased snowfall. Whether the one effect will cancel out the other in terms of sea-level rise has yet to be determined with confidence.

Mountain glaciers

Whereas the effect of global warming on the ice sheets is uncertain, but might be neutral in the short term, the same cannot be said for the world's mountain glaciers. Ever since the Little Ice Age of the eighteenth and nineteenth centuries, most of the world's mountain glaciers have been receding. The recession has not been uniform, however, and the general trend in many areas has been interrupted by re-advances in response to decadal changes in temperature and precipitation.

Even within the last decade, many glaciers in Norway have been advancing, in response to increased winter precipitation, associated with more vigorous depressions. Two famous glaciers in New Zealand, the Fox and Franz Josef, advanced strongly throughout the 1980s and 1990s. In the Alps, the 1970s saw lower temperatures, and up to half the glaciers in Switzerland began to advance. In Alaska a number of tidewater glaciers bucked the general trend in the latter part of the twentieth century by creating their own sediment platforms over which to advance. However, the general pattern has been one of major recession to the extent that many glaciers lost half their mass in the twentieth century, and some even disappeared altogether.

Spectacular glacial retreat is causing the loss of scenic beauty spots in tropical areas. This beautiful valley glacier in the Cordillera Huayhuash, seen here in 1980, descending from Nevado Yerupaja has receded so much that the tongue no longer reaches the water. No more 'tropical icebergs' are calved into this moraine-dammed lake.

Now, in the early part of the twenty-first century, recession is strong in many parts of the world, and apparently accelerating. Calculations by an American team in 2002, which has collected data from all the world's mountain glaciers, indicate that the rate of ice loss has more than doubled since 1988. Of the world's mountain glaciers, those in Alaska and northern Canada have contributed 0.32 millimetres to the raising of the sea level per year in the last decade, which represents about half the global rate of ice loss.

Most tidewater glaciers in the fjords of southern Alaska have been receding rapidly over the past century. However, like Surprise Glacier in Harriman Fiord, a branch of College Fiord, they still flow vigorously, discharging large volumes of icebergs into the fjord.

Overall effect of sea-level rise

The figures given by the IPCC for projected sea-level rise this century are tabulated opposite. Although they reveal a large element of uncertainty, the overall effect is for substantial sea-level rise even for the best-case scenario, whilst the worst case could have dire consequences around the world, especially when combined with a rise attributable to thermal expansion of the oceans.

A one-metre sea-level rise in Bangladesh, for example, would inundate half the country and displace 100 million people. Many low-lying islands such as Kiribati, southwest of Hawaii in the Pacific, and the Seychelles in the Indian Ocean would be flooded. Costly reinforcement of sea defences in developed countries would be necessary to protect low-lying areas such as the Fens in eastern England, most of Holland, parts of northern Germany, the coastal fringe of Florida and the Houston area of Texas.

Projected global average sea-level rise from 1990 to 2100. Positive values indicate rise; negative values a lowering according to the Intergovernmental Panel on Climate Change[a]

Source of sea-level rise	Metres of sea-level rise, 1990–2100
Antarctica	−0.17 to +0.02
Greenland	−0.02 to +0.09
Mountain glaciers and ice caps	+0.01 to +0.23
Thermal expansion	+0.11 to +0.43

[a] (Houghton, J. T., Ding, Y., Griggs, D. J., Noguer, M., van der Linden, P. J., Dai, X., Maskell, K., Johnson, C. A. (eds.) *Climate Change 2001: The Scientific Basis*, Contribution of Working Group I to the Third Assessment Report of the Intergovernmental Panel on Climate Change. Cambridge: Cambridge University Press.)

With continued global warming in the longer term, over several centuries, sea-level changes are likely to be even more dramatic. The large-scale ice sheets may well have a delayed response to warming, so it is hard to predict when they might disappear. If, however, the entire Antarctic Ice Sheet were to melt, a volume of about 30 million cubic kilometres of water would be added to the oceans. This is equivalent to a sea-level rise of about 60 to 70 metres, and would result in many of the world's major cities, such as London, New York, Buenos Aires, Calcutta, Shanghai and Tokyo, being submerged. Although this is unlikely to happen in the foreseeable future, despite the spread of scare stories in recent years, it does show that even small percentage volume changes of the Antarctic Ice Sheet could have considerable effects. However, it is beyond AD 2100 that the real impact of Antarctic Ice Sheet melting will kick in. Similarly, the 6.5 metres of sea-level rise from melting of the Greenland Ice Sheet will probably be delayed until after 2100.

At present we seem to be heading towards irreversible damage to the Earth's ecosystem as a result of global warming, and the demise of glaciers is just one manifestation of this. Even though some governments have accepted the validity of bodies like the IPCC, there are others that choose to ignore the warnings. Yet to slow down the climatic warming caused by human civilization will require international agreement from all polluting countries to cut down emissions, and an acceptance by people in developed countries in particular that we cannot maintain our extravagant lifestyles. Investment in education concerning environmental matters is necessary to slow or, preferably, reverse this trend, for we are currently running extreme environmental risks with little understanding of what the likely outcome will be.

Impact of loss of water resources from glaciers

Water from mountain glaciers is a valuable resource that in some areas is rapidly disappearing. Tropical countries such as Bolivia and Peru currently benefit from extensive glacierization in the high Andes, and the millions of people in their capital cities of La Paz and Lima respectively rely on melt from these sources. However, all tropical glaciers are receding at such a rate that some scientists have predicted that most will disappear by 2020. The consequences of this trend are extremely severe, as few alternative sources of water exist. Agricultural areas also rely on these Andean glaciers for irrigation during the dry season, so any loss of water supply will make these areas unviable. In other areas, such as central Asia and Argentina, recession could be accompanied by substantially increased discharge in summer, thereby increasing the risk of flooding and landslides.

In the more developed countries of mid-latitudes, glacier water is not only used for irrigation, but also for the generation of hydro-electricity. Loss of glacier mass will ultimately reduce the run-off to the reservoirs that feed the turbines, and with it the capacity of countries such as Norway and those in the Alps to generate 'green' energy. Overall, some scientists estimate that, on current trends, up to a quarter of mountain glacier mass will be lost by 2050, and up to a half by 2100.

Opposite. The Southern Patagonian Icefield in Chile feeds numerous outlet glaciers. Some of those flowing east, like Glaciar Grey, terminate in deep lakes that were carved out by ice during the last ice age. The snout of this and similar glaciers terminates in the beautiful forests of southern beech (*Nothofagus*).

The scale of these glacierized peaks in the Karakorum Mountains is evident from a flight from London to Beijing. The pyramidal peak is K2 (8611 metres), the world's second highest peak, and is a typical 'horn'. Debris-covered glaciers, fed by ice and rock avalanches, descend into the dark valleys below.

Increased potential for glacier disasters

In high mountain regions, notably the Himalaya and tropical Andes, glacier recession is likely to generate an increased risk of glacier disaster. The tongues of many glaciers in these regions are debris-covered, and so they have been protected from the rapid recession experienced by clean glaciers. As global warming gathers pace, these glaciers are becoming more vulnerable to melting. Now, many glaciers in the Himalaya and Andes are receding back from their Little Ice Age moraines, with little prospect of regaining mass. Lakes are forming between the moraine and the glacier, and potentially could burst. Several major disasters occurred in the twentieth century, and it is predicted that increasing numbers of glaciers will generate such hazardous lakes. At the same time, glacier recession exposes unstable rock faces and debris slopes, increasing the risk of landslides. If a landslide were to fall into a glacial lake, then the moraine dam could fail catastrophically, as has happened several times in the past. Indeed, landslides are an increasing risk in all mountain areas with glaciers, since glaciers over-steepen the valley sides that then are prone to failure as the glacier recedes.

Impact of glacier loss on tourism

Glaciers are a major tourist attraction in many parts of the world. Lifts have been built on them so that skiers can continue their activities outside the main skiing season. Loss of glaciers will have a severe impact on some high-altitude resorts, as indeed has already happened in the Alps. Mountaineers use glaciers as highways to their peaks, and recession in many cases will make access more difficult. The sightseer has currently many vantage points from which to admire these beautiful natural phenomena. As they recede, glaciers become increasingly dirty, and leave behind vast piles of vegetation-free rubble that is much less attractive to the tourist.

If ice ages were to continue their cyclical pattern, then areas like Loch Torridon in the Northwest Highland of Scotland, which bear a strong glacial imprint, will once again be inundated by glacier ice in several thousand years time. It remains to be seen if human-induced climatic warming will ultimately pre-empt such a development.

Will ice sheets come back?

Looking further ahead, the real question is: 'will the short-term effects of human civilization's disruption to climate be irreversible?' If not, it is possible that glaciers will have the last word. We are living in an interglacial period within a long-running glacial era that has seen ice sheets wax and wane on a large scale. A mere 20 000 years ago, ice sheets covered a third of the land on the Earth. This was part of a recurring cycle of ice ages and interglacial periods, and the natural trend would be to see another ice age. Thus, in a few thousand years ice sheets may once again extend towards mid-latitudes, across the greater part of northwest Europe, North America and elsewhere.

Endnote

In writing this book we have attempted to explain in simple terms how glaciers work, and how they have an impact on landscape and human civilization. However, first and foremost we have tried to convey the beauty and fascination of glaciers, based on our personal experiences. Photographs only partly illustrate these attributes, and glaciers can only be fully appreciated by actually visiting them. If this book does nothing else, we hope that it will implant in the reader a desire to visit glaciers and glacial landscapes, and gain a deeper appreciation of this magnificent but vulnerable world of snow and ice.

Opposite. Grounded icebergs form a mirage off the coast near Davis Station in East Antarctica at sunset. If the major ice sheets were to collapse catastrophically, as some scientists have predicted, sights like this will become commonplace.

Glossary

Ablation The process of wastage of snow or ice by melting, sublimation and calving.

Ablation area/zone That part of a glacier's surface, usually at lower elevations, over which ablation exceeds accumulation.

Ablation valley The name given to the subsidiary valley that forms between the crest of a lateral moraine (q.v.) and the rocky valley side.

Accumulation The process of building-up of a pack of snow, refrozen slush, meltwater and firn (q.v.). Net accumulation for one year is the material left over at the end of the melt-season.

Accumulation area That part of a glacier's surface, usually at higher elevations, on which there is net accumulation of snow, which subsequently turns into firn and then glacier ice.

Aluvión (from Peruvian Spanish) A catastrophic flood or debris flow, commonly resulting from the failure of a moraine-dammed lake.

Andesite A viscous (sticky) silica-rich lava extruded during an explosive volcanic eruption at destructive plate boundaries, as around the Pacific Rim.

Aquifer An underground reservoir of water, usually found in porous rocks and sediments, including glacial deposits.

Areal scouring Large-scale erosion of bedrock in lowland areas by an ice sheet.

Arête (from French) A sharp, narrow, often pinnacled ridge, formed as a result of glacial erosion from both sides.

Basal debris Rock fragments and ground-up bedrock incorporated into the base of a glacier.

Basal ice layer The layer of ice at the bed of a glacier that is the product of melting and refreezing (regelation, q.v.). It is strongly layered, sheared and incorporates a variable amount of debris.

Basal sliding The sliding of a glacier over bedrock, a process usually facilitated by the lubricating effect of meltwater.

Basalt A highly mobile lava extruded typically during a fissure eruption at a constructive tectonic plate boundary, e.g. in Iceland. A dark grey rock rich in iron and magnesium.

Basket-of-eggs topography Extensive low-lying areas covered by small elongated hills called drumlins (q.v.).

Bergschrund (from German) An irregular crevasse, usually running across an ice slope in the accumulation area (q.v.), where active glacier ice pulls away from ice adhering to the steep mountainside.

Bergy bit A piece of floating glacier ice up to several metres across, commonly deriving from the disintegration of an iceberg.

Boulder clay An English term for till (q.v.), no longer favoured by glacial geologists.

Braided stream A relatively shallow stream that has many branches that commonly recombine and migrate across a valley floor. Braided streams typically form downstream of a glacier.

Breached watershed A short, glacially eroded valley, linking two major valleys across a mountain divide.

Calving The process of detachment of blocks of ice from a glacier into water.

Chattermarks A group of crescent-shaped friction cracks on bedrock, formed by the juddering effect of moving ice.

Cirque (from French) An armchair-shaped hollow with steep sides and back wall, formed as a result of glacial erosion high on a mountainside, and often containing a rock basin with a tarn (q.v.) (cf. Corrie, Cwm).

Cirque glacier A glacier occupying a cirque.

Col (from French) A high-level pass formed by glacial breaching of an arête (q.v.) or mountain mass.

Cold glacier A glacier in which the bulk of the ice is below the pressure melting point and therefore frozen to the bed.

Cold ice Ice which is below the pressure melting point, and therefore dry.

Compressive flow The character of ice flow where a glacier is slowing down and the ice is being compressed and thickened in a longitudinal direction.

Conduit A drainage tunnel within or at the bed of a glacier.

Corrie (from Gaelic *coire*) A British term for cirque (q.v.).

Crag-and-tail A glacially eroded rocky hill with a tail of till formed down-glacier of it.

Crescentic gouge A crescent-shaped scallop, usually several centimetres across, formed as a result of bedrock fracture under moving ice.

Crevasse A deep V-shaped cleft formed in the upper brittle part of a glacier as a result of the fracture of ice undergoing extension.

Crevasse traces Long veins of clear ice a few centimetres wide, formed as a result of fracture and recrystallization of ice under tension without separation of the two walls; these structures commonly form parallel to open crevasses and extend into them. Thicker veins of clear ice resulting from the freezing of standing water in open crevasses are also called crevasse traces.

Cryoconite hole A small cylindrical hole on the surface of a glacier, formed by small patches of debris that absorb more radiation than the surrounding ice, and melt downwards at a faster rate.

Cryosphere A general term to embrace all ice-covered areas on planet Earth, including ice on land and on the sea.

Cwm The Welsh term for cirque (q.v.), also sometimes used more generally outside Wales.

Débacle (from the French) A catastrophic flood or debris flow, commonly resulting from the failure of a moraine- or ice-dammed lake.

Dirt cone A thin veneer of debris draping a cone of ice up to several metres high, formed because the debris has retarded ablation under it.

Drift A nineteenth century term, still in use, to describe all unconsolidated deposits associated with glaciers, glacial meltwater and icebergs.

Drumlin (from Gaelic) A streamlined hillock, commonly elongated parallel to the former ice flow direction, composed of glacial debris, and sometimes having a bedrock core; formed beneath an actively flowing glacier.

Eccentricity Change of shape of the Earth's elliptical orbit on cycles of 100 000 and 400 000 years. One of three astronomical variables that affect the amount of solar radiation received at the Earth's surface.

Ejecta The explosive products of an erupting volcano.

Englacial debris Debris dispersed throughout the interior of a glacier. It originates either in surface debris that is buried in the accumulation area or falls into crevasses, or in basal debris that is raised by thrusting or folding.

Englacial stream A meltwater stream that has penetrated below the
   surface of a glacier and is making its way towards the bed or sides.

Equilibrium line/zone The line or zone on a glacier's surface where a
   year's ablation balances a year's accumulation (cf. Firn line). It is
   determined at the end of the ablation season, and commonly occurs at
   the boundary between superimposed ice (q.v.) and glacier ice.

Erratic A boulder or large block of bedrock that is being, or has been,
   transported away from its source by a glacier.

Esker (from Gaelic) A long, commonly sinuous ridge of sand and gravel,
   deposited by a stream in a subglacial tunnel.

Extending flow The character of ice flow where a glacier is accelerating
   and the ice is being stretched and thinned in a longitudinal
   direction.

Fault A displacement in a glacier formed by ice fracturing without its
   walls separating. It can be recognized by the discordance of layers in
   the ice on either side of the fracture.

Firn (from German) Dense, old snow in which the crystals are partly
   joined together, but in which the air pockets still communicate with
   each other.

Firn line The line on a glacier that separates bare ice from snow at the end
   of the ablation season.

Fjord (from Norwegian; **Fiord** in North America and New Zealand) A
   long, narrow arm of the sea, formed as a result of erosion by a valley
   glacier.

Fold Layers of ice that have been deformed into curved forms by flow at
   depth in a glacier.

Foliation Groups of closely spaced, often discontinuous, layers of coarse
   bubbly, coarse clear and fine-grained ice, formed as a result of shear
   or of compression at depth within a glacier.

Frictional heat The heat generated from the rubbing effect as a glacier,
   especially if debris-laden, moves over its bed.

Geothermal heat The heat output from the Earth's surface. This affects
   glaciers especially in the polar regions, by warming the basal zone to
   the pressure melting point.

Glacial lake outburst flood (GLOF) During a period of glacier recession
   back from a terminal moraine a lake may form, impounded by an
   unstable pile of debris and buried ice. Catastrophic failure of the

moraine will result in a devastating flood. Usually associated with high mountain regions such as the Andes and Himalaya.

Glacial outwash deposits Sediment produced and reworked by glacial meltwater streams.

Glacial period/glaciation A period of time when large areas of the Earth (including present temperate latitudes) were covered by ice. Numerous glacial periods have occurred within the last few million years, and are separated by interglacial periods (q.v.). Glacial periods have also occurred sporadically throughout geological time.

Glacial trough A glaciated valley or fjord (q.v.), often characterized by steep sides and a flat bottom, with multiple basins, resulting primarily from abrasion by strongly channelled ice.

Glaciated The character of land that was once covered by glacier ice in the past (cf. Glacierized).

Glacier A mass of ice, irrespective of size, derived largely from snow, and continuously moving from higher to lower ground.

Glacier ice Any ice in, or originating from, a glacier, whether on land or floating on the sea as icebergs.

Glacier karst Debris-covered stagnant ice, sometimes found at the snout of a retreating glacier, with numerous lake-bearing caverns and tunnels.

Glacier milk Meltwater from glaciers, which commonly has a milky appearance as a result of suspended fine sediment.

Glacier portal The entrance to an open tunnel, formed where subglacial meltwater emerges at the glacier snout or terminus.

Glacier sole The lower few metres of a (usually sliding) glacier that are rich in debris picked up from the bed.

Glacier table A boulder perched on a pedestal of ice; the boulder protecting the ice from ablation during sunny weather.

Glacier tongue (or **ice tongue**) An unconstrained, floating extension of an ice stream or valley glacier, projecting into the sea.

Glacierized The character of land currently covered by glacier ice (cf. Glaciated).

Groove A glacial abrasional form, with striated (q.v.) sides and base, orientated parallel to the ice-flow direction, and commonly several metres wide and deep.

Grounding line/zone The line or zone at which an ice mass enters the sea or a lake and begins to float, e.g. in the inner part of an ice shelf (q.v.) or an ice stream (q.v.).

Growler A piece of glacier ice almost awash, up to a few metres across, but generally smaller than a bergy bit (q.v.).

Hanging glacier A glacier that spills out from a high level cirque (q.v.) or clings to a steep mountainside.

Hanging valley A tributary valley whose mouth ends abruptly part way up the side of a trunk valley, as a result of the greater amount of glacial down-cutting of the latter.

Highland icefield A near-continuous stretch of glacier ice, but with an irregular surface that follows approximately the contours of the underlying bedrock, and which is punctuated by nunataks (q.v.).

Horn A steep-sided, pyramid-shaped peak, formed as a result of the backward erosion of cirque glaciers on three or more sides.

Hyaloclastite Lava (usually basalt) which has become fragmented by quenching in water. In a glacial context this fragmentation takes place as a result of sudden melting when lavas are extruded subglacially.

Ice age A period of time when large ice sheets (q.v.) extend from the polar regions into temperate latitudes. The term is sometimes used synonymously with 'glacial period' (q.v.), or embraces several such periods to define a major phase in Earth's climatic history.

Ice apron A steep mass of smooth ice, commonly the source of ice avalanches, that adheres to steep rock near the summits of high peaks.

Ice cap A dome-shaped mass of glacier ice, usually situated in a highland area, and generally defined as covering up to 50 000 square kilometres (cf. Ice sheet).

Ice cliff (ice wall) A vertical face of ice, normally formed where a glacier terminates in the sea, or is undercut by streams. These terms are also used more specifically for the face that forms at the seaward margin of an ice sheet or ice cap, and that rests on bedrock at or below sea level.

Ice-dammed lake A temporary lake, dammed by a glacier or where two glaciers merge. Prone to seasonal catastrophic drainage.

Ice sheet A mass of ice and snow of considerable thickness and covering an area of more than 50 000 square kilometres.

Ice shelf A large slab of ice floating on the sea, but remaining attached to and largely fed by land-derived ice.

Ice ship A pinnacle of ice, shaped like a triangular sail, typically several metres high, formed as a result of differential ablation under strong solar radiation in low latitudes.

Ice stream Part of an ice sheet or ice cap in which the ice flows more rapidly, and not necessarily in the same direction as the surrounding ice. Zones of strongly sheared, crevassed ice often define the margins (cf. Shear zone).

Ice tongue (*see* Glacier tongue)

Iceberg A piece of ice of the order of tens of metres to many kilometres across that has been shed by a glacier terminating in the sea or a lake.

Icefall A steep reach in a glacier where the ice accelerates and becomes heavily crevassed, commonly over a bedrock step.

Interglacial period A period of time, such as the present day, when ice still covers parts of the Earth's surface, but has receded to the polar or high mountain regions.

Internal deformation That component of glacier flow that is the result of the deformation of glacier ice under the influence of accumulated snow and firn and of gravity.

Isotopes Varieties of elements, all with identical chemical properties, but not precisely the same physical ones.

Jökulhlaup (from Icelandic) A sudden and often catastrophic outburst of water from a glacier during a volcanic eruption. The term is also used to describe when an ice-dammed lake bursts or an internal water pocket escapes, resulting in flooding.

Kame (from Gaelic) A steep-sided hill of sand and gravel deposited by glacial streams adjacent to a glacier margin.

Kame terrace A flat or gently sloping plain, deposited by a stream that flowed towards or along the margin of a glacier, but that was left above the hillside when the ice receded.

Kettle (or **kettle hole**) A self-contained bowl-shaped depression within an area covered by glacial stream deposits, often containing a pond. A kettle forms from the burial of a mass of glacier ice by stream sediment and its subsequent melting.

Kinematic wave The means whereby mass-balance changes are propagated down-glacier. The wave has a constant discharge and moves faster than the ice itself. Kinematic waves are visible as bulges on the ice surface, and are especially prominent in surging glaciers. Once the wave reaches the snout, the glacier is able to advance.

Knock-and-lochan topography (from Gaelic) Rough, ice-abraded, low-level landscape, comprising small hills of exposed bedrock, and rock basins with small lakes and bogs.

Lahar Debris-flow consisting primarily of volcanic ash and lava boulders. Heavy rain and/or melting snow and ice during a volcanic eruption mixes with the loose deposits and forms fast-moving tongues of slurry.

Little Ice Age The period of time that led to the expansion of valley and cirque glaciers world-wide, with their maximum extents being attained in about AD 1700–1850 in many temperate regions and around 1900 in Arctic regions.

Loess Wind-blown silt, often derived from the fine-grained material deposited on outwash plains in front of glaciers, and carried long distances away from the source.

Mass balance (or **mass budget**) A year-by-year measure of the state of health of a glacier, reflecting the balance between accumulation and ablation. A glacier with a positive mass balance in a particular year gained more mass through accumulation than was lost through ablation; the reverse is true for negative mass balance.

Moraine Distinct ridge or mound of debris laid down directly by a glacier or pushed up by it. The material is mainly till, but fluvial, lake or marine sediments may also be involved. Longitudinal moraines (parallel to the valley) include a **lateral moraine** which forms along the side of a glacier; a **medial moraine** occurring on the surface where two streams of ice merge; and a **fluted moraine** which forms a series of ridges beneath the ice, parallel to flow. Transverse moraines include a **terminal** (or **end**) **moraine** which forms at the furthest limit reached by the ice, a **recessional moraine** which represents a stationary phase during otherwise general retreat, and a set of **annual moraines** representing a series of minor winter readvances during a general retreat. A **push moraine** is a more complex form that develops especially in front of a polythermal glacier during a period of advance. Disorganized mounds of glacially deposited debris are referred to as **hummocky moraines**.

Moraine-dammed lake A lake formed as a glacier recedes from its terminal moraine, the moraine acting as an unstable dam (*see* Glacial lake outburst flood).

Moulin (from French) A water-worn pothole formed where a surface meltstream exploits a weakness in the ice. Many moulins are cylindrical, several metres across, and extend down to the glacier bed, although often in a series of steps.

Nunatak (from Inuit) An island of bedrock or mountain projecting above the surface of an ice sheet or highland icefield.

Nye channel (named after a British physicist) A channel cut into bedrock by subglacial meltwater under high pressure. Usually less than one metre wide. Commonly deeper than it is wide.

Obliquity of the ecliptic Variations in the Earth's axis of tilt, which take place on a 41 000-year cycle. One of three astronomical variables that affect the amount of solar radiation received at the Earth's surface.

Ogives Arcuate bands or waves, with their apices pointing down-glacier, that develop in an ice-fall. Alternating light and dark bands are called banded ogives or Forbes' bands. Each pair of bands or one wave and trough is believed to represent a year's movement through the icefall.

Outlet glacier A glacier tongue that originates from an **ice sheet, ice cap** or **highland icefield**, i.e. it has no clearly defined **accumulation area**.

Outwash plain A flat spread of debris deposited by meltwater streams emanating from a glacier (cf. Sandur).

Overspill (overflow) channel A channel cut through a hill or ridge by meltwater, as a result of ice damming a lake.

Oxygen isotope analysis The technique of examining the ratios of the heavy and light isotopes of oxygen in snow, ice or marine sediment, that can indirectly provide a climatic records as well as an estimate of the volume of glacier ice on Earth.

p-forms (or plastically moulded forms) Smooth rounded forms of various types cut into bedrock by the combined erosive power of ice, meltwater and subglacial sediment under high pressure.

Permafrost Ground that remains permanently frozen. It may be hundreds of metres thick with only the top few metres thawing out in summer.

Piedmont glacier A glacier that spreads out as a wide lobe as it leaves a narrow mountain valley to enter a wider valley or a plain.

Polythermal glacier A thermally complex glacier with both warm and cold ice (q.v.). Typically, warm ice occurs where the ice is thickest as a result of geothermal heating.

Portal The open archway that develops when a meltwater stream emerges at the snout of a glacier.

Precession of the equinoxes Changes in the rotation of the Earth's spin axis as the planet orbits the Sun on a 19000- to 23000-year cycle. One of three astronomical variables that affect the amount of solar radiation received at the Earth's surface.

Pressure melting point The temperature at which ice melts under a specific pressure. Pressure lowers the melting point to below 0 °C beneath a glacier.

Proglacial lake A lake developed immediately in front of the glacier, commonly bordered by the mounds of unconsolidated deposits that characterize the terminal zone of a glacier.

Pyroclastic deposits The products of explosive volcanic eruptions, falling out from the eruption column and from fast-flowing dense clouds of ash, or as debris flowing down-slope especially when mixed with water and melting ice.

Quiescent phase The period in which a surge-type glacier is slow-moving or stagnant. This period is typically decades long in contrast to the surge phase that may last only a few months or years.

Randkluft (from German) The narrow gap that develops between a rock face and steep firn and ice at the head of a glacier.

Regelation ice Ice which is formed from meltwater as a result of changes of pressure beneath a glacier.

Rejuvenated (or **regenerated**) **glacier** A glacier which develops from ice avalanche material beneath a rock cliff.

Reservoir area The upper reaches of a surge-type glacier where ice slowly builds up without being transferred down-valley until a surge takes place.

Riegel (from German) A rock barrier that extends across a glaciated valley, usually comprising harder rock, and often having a smooth up-valley facing slope and a rough down-valley facing slope.

Rill The small channel produced by running water on the surface of a glacier or sediment, usually a few centimetres across and ephemeral.

Roche moutonnée (from French) A rocky hillock with a gently inclined, smooth up-valley facing slope resulting from glacial abrasion, and a steep, rough down-valley facing slope resulting from glacial plucking.

Rock basin A lake- or sea-filled bedrock depression carved out by a glacier.

Rock flour Bedrock that has been pulverized at the bed of a glacier into clay- and silt-sized particles. It commonly is carried in suspension in glacial meltwater streams, which consequently take on a milky appearance (cf. Glacier milk).

Run-out distance The distance an ice avalanche travels from its source.

Sandur (plur. **Sandar**) (from Icelandic) Extensive flat plain of sand and gravel with braided streams of glacial meltwater flowing across it. Sandar are usually not bounded by valley walls and commonly form in coastal areas.

Sea ice Ice that forms by the freezing of sea water. (cf. Ice shelves and icebergs that also float on the sea and are derived from glacier ice on land.)

Sea loch The Scottish term for a fjord (q.v.).

Sedimentary stratification The annual layering that forms from the accumulation of snow, and that is preserved in firn and sometimes in glacier ice.

Sérac (from French) A tower of unstable ice that forms between crevasses, often in icefalls or other regions of accelerated glacier flow.

Shear zone A zone of severe deformation, especially where a fast-flowing ice stream (q.v.) moves past relatively slow-moving ice. The deformation is characterized especially by intense crevassing.

Sill A submarine barrier of rock or moraine that occurs at the mouth of, or between rock basins in, a fjord.

Slush flow (slush avalanche) A fast-flowing mass of water-saturated snow, most commonly occurring in early summer when melting is at its peak.

Snout The lower part of the ablation area of a valley glacier, often shaped like the snout of an animal.

Snow swamp An area of saturated snow lying on glacier ice.

Snowball Earth The name given to the hypothesis that the Earth was almost completely covered by glacier ice in Late Proterozoic time (1000 to 570 million years ago).

Sole *See* Glacier sole.

Stapi (from Icelandic; also named **Tuya**) A subglacially erupted volcano, characterized by steep flanks and a flat or slightly domed summit. Named after a volcano in Iceland, no longer enveloped in ice.

Strain The amount by which an object becomes deformed as a result of stress.

Stress The measure of how hard a material (such as glacier ice) is being pushed or pulled as a result of an external force.

Striae or **striations** Linear, fine scratches formed by the abrasive effect of debris-rich ice sliding over bedrock. Intersecting sets of striae are formed as stones are rotated or if the direction of the ice flow over bedrock changes.

Striated The scratched state of bedrock or stone surfaces after the ice has moved over them.

Subglacial debris Debris which has been released from ice at the base of a glacier. Individual stones usually show signs of rounding as a result of abrasion at the contact between ice and bedrock.

Subglacial meltwater channel A steep, often vertically sided valley cut into bedrock by a subglacial stream under high pressure. It may have a profile with uphill sections, since a sufficient head of water in the glacier can force it to flow up slope.

Subglacial stream A stream that flows beneath a glacier, and which usually cuts into the ice above to form a tunnel.

Superimposed ice Ice that forms as a result of the freezing of water-saturated snow. It commonly forms at the surface of a glacier between the equilibrium line and the firn line, and gives the glacier additional mass.

Supraglacial debris Debris which is carried on the surface of a glacier. It is normally derived from rock-falls and tends to be angular in character.

Supraglacial stream A stream that flows over the surface of a glacier. Most supraglacial streams descend via moulins (q.v.) into the depths or base of a glacier.

Surge front The zone of intense compression between surging ice and non-surging ice. This is commonly marked by a bulge and a transition from heavily crevassed to crevasse-free ice. The surge front rapidly moves through the glacier, and if it reaches the snout, the glacier advances.

Surge phase A short-lived phase of accelerated glacier flow during which the surface becomes broken up into a maze of crevasses. Surges are often periodic and are separated by longer periods of relative inactivity or even stagnation (cf. Quiescent phase).

Tabular iceberg A flat-topped iceberg that has become detached from an ice shelf, ice tongue or floating tidewater glacier; typically several kilometres long.

Tarn A small lake occupying a hollow eroded out by ice or dammed by a moraine; especially common in cirques.

Temperate glacier A glacier whose temperature is at the pressure melting point throughout, except for a cold wave of limited penetration that occurs in winter.

Tephra Ash and coarse fragments thrown out by a volcano during an eruption. In glacierized areas these layers are preserved in glaciers, deforming slowly during ice flow.

Thermal regime That state of a glacier as determined by its temperature distribution.

Thrust A low-angle fault, usually formed where the ice is under compression. Thrusts commonly extend from the bed and are associated with debris and overturned folds.

Tidewater glacier A glacier that terminates in the sea. Some writers restrict the term to a glacier with its terminus resting on the sea floor.

Till A mixture of mud, sand and gravel-sized material deposited directly from glacier ice. The principal types of till are **basal**, deposited beneath a glacier, and embracing **lodgement till** (plastered on the bed) and **meltout till** (released from slow-moving or stagnant ice); and **supraglacial meltout till**, let down onto the substrate from the glacier surface.

Tillite The hard rock equivalent of till.

Tongue That part of a valley glacier that extends below the firn line.

Trim-line A sharp line on a hillside marking the boundary between well-vegetated terrain that has remained ice-free for a considerable time and poorly vegetated terrain that until relatively recently lay under glacier ice. In many areas the most prominent trim-lines date from the Little Ice Age (q.v.).

Trough-mouth fan A large-scale arcuate accumulation of sediment, built out from the edge of the continental shelf when a glacier reached this position and deposited its load at the break of slope. Typically many tens of kilometres across.

Tundra The zone of shrubs and other small plants that grow mainly on top of the permafrost in Arctic regions north of the tree line.

Tuya (*see* Stapi)

Unconformity A discontinuity in the annual layering in firn or ice, resulting from a period when ablation cut across successive layers.

Valley glacier A glacier bounded by the walls of a valley, and descending from high mountains, from an ice cap on a plateau, or from an ice sheet.

Warm glacier (*see* Temperate glacier)

Warm ice Ice that is at melting point. The temperature may be slightly below 0 °C at the base of a glacier where the ice is under high pressure.

Water table The level up to which a glacier is saturated with water. Applying principally to temperate glaciers, the water flows along ice crystal boundaries and up into cavities such as moulins and conduits (q.v.) until it reaches a stable level. The water table fluctuates on daily and seasonal cycles.

Wet snow zone The zone of slush that forms in the vicinity of the snow-line. It migrates up-glacier during the course of the melt-season.

Whaleback A smooth, scratched, glacially eroded bedrock knoll several metres high, and resembling a whale in profile.

Select bibliography

This list gives a selection of recent books concerning glaciers and their products, including those suitable for the wider public as well as for secondary school students, and university undergraduates and postgraduates.

General interest books

Gordon, J. *Glaciers*. Grantown-on-Spey, Scotland: Colin Baxter Photography, 2001.
  A short paperback with excellent colour photographs, supported by a brief simple text.
Hambrey, M. J. & Alean, J. *Glaciers*. Cambridge: Cambridge University Press, 1992.
  Shorter first edition of the present volume, aimed at the educated layperson, with a strong emphasis on black-and-white and colour photographs (mostly different from those contained herein).
Post, A. & LaChapelle, E. R. *Glacier Ice*. Seattle: University of Washington Press; Cambridge: The International Glaciological Society, 2000.
  Small-scale and updated version of the classic 1971 large-format volume of the same title, now a collectors' item. Stunning black-and-white pictures, focusing particularly on Alaska and western North America from the air. An informative text explains the workings of glaciers.
Sharp, R. P. *Living Ice*. Cambridge: Cambridge University Press, 1988.
  An informative volume, written in an engaging non-technical style for the educated layperson, supported by a mix of colour and black-and-white photographs. Strong North American emphasis.

Text books

Benn, D. I. & Evans, D. J. A. *Glaciers and Glaciation*. London: Arnold, 1999.
  A university-level textbook, covering the broad spectrum of glaciers and glacial

geology in considerable detail. With its comprehensive list of references, this book represents the 'state-of-the-art' in the field, and is a must for undergraduates.

Bennett, M. R. & Glasser, N. F. *Glacial Geology: Ice Sheets and Landforms.* Chichester: John Wiley & Sons, 1996.

A well-written textbook for university undergraduates, covering the broad spectrum of the subject. A unique feature is the inclusion of 'boxes' which give in-depth summaries of interesting scientific papers.

Hambrey, M. J. *Glacial Environments.* Vancouver: University of British Columbia Press; London: UCL Press, 1994.

A well-illustrated undergraduate-level textbook, with numerous black-and-white photographs. There is a strong emphasis on sediments and landforms, and it covers the marine environment particularly thoroughly.

Knight, P. J. *Glaciers.* Cheltenham: Stanley Thornes (Publishers) Ltd., 1999.

An undergraduate/postgraduate-level textbook, covering the principles of glacier behaviour thoroughly for people of limited mathematical training.

Menzies, J., ed. *Modern Glacial Environments: Processes, Dynamics and Sediments.* Oxford: Butterworth-Heinemann, 1995.

A collection of articles on different aspects of glacial environments by experts in each field, mainly for the researcher and advanced undergraduate. Good lead-in to additional literature.

Nesje, A. & Dahl, S. O. *Glaciers and Environmental Change.* London: Arnold, 2000.

A relatively short, but wide-ranging book dealing with the essentials of glaciology and glacial geology within the context of global environmental change.

Paterson, W. S. B. *Physics of Glaciers.* 3rd edn. Oxford: Pergamon, 1994.

The best up-to-date single reference book dealing with the physical principles of glacier formation and behaviour. Although an understanding of high-level mathematics and physics is desirable, other readers will find much of interest in this volume.

Siegert, M. J. *Ice Sheets and Late Quaternary Environmental Change.* Chichester: Wiley, 2001.

A comprehensive review of present and former ice sheets, their mode of formation and how modelling can be used to predict their dimensions. Suitable for undergraduates, but also of value for researchers.

Glacier monitoring and safety

Kaser, G., Fountain, A. & Jansson, P. *A Manual for Monitoring the Mass Balance of Mountain Glaciers.* International Hydrological Programme. IHP-VI Technical Documents in Hydrology No. 59, UNESCO, Paris, 2003. Accessed via: http://unesdoc.unesco.org/images/0012/001295/129593e.pdf.
This official technical report includes a comprehensive account of how to travel safely on mountain glaciers, including crevasse rescue. Other mountaineering manuals generally have sections on glacier travel.

Websites

Typing 'Glaciers' into the Google search engine yielded 450 000 entries in August 2003. To make a selection from these would be invidious, but sites operated by governmental bodies and universities are particularly useful. Readers are encouraged to search the Web for themselves, as it contains a mine of information. More about the authors' interests can be found as follows:

Jürg Alean: http://www.stromboli.net

Michael Hambrey: http://www.aber.ac.uk/glaciology/

Location index

Subject Index

Numbers in bold refer to illustrations